MW00843845

Electrifying the Piedmont Carolinas

The Duke Power Company, 1904–1997

Electrifying the Piedmont Carolinas

The Duke Power Company, 1904–1997

Robert F. Durden

Carolina Academic Press
Durham, North Carolina

ISBN 0-89089-743-3
LCCN 2001087075

Carolina Academic Press
700 Kent Street
Durham, NC 27701
Telephone (919) 489-7486
Fax (919) 493-5668
www.cap-press.com

All photographs are from the Duke Power Archives.

Printed in the United States of America

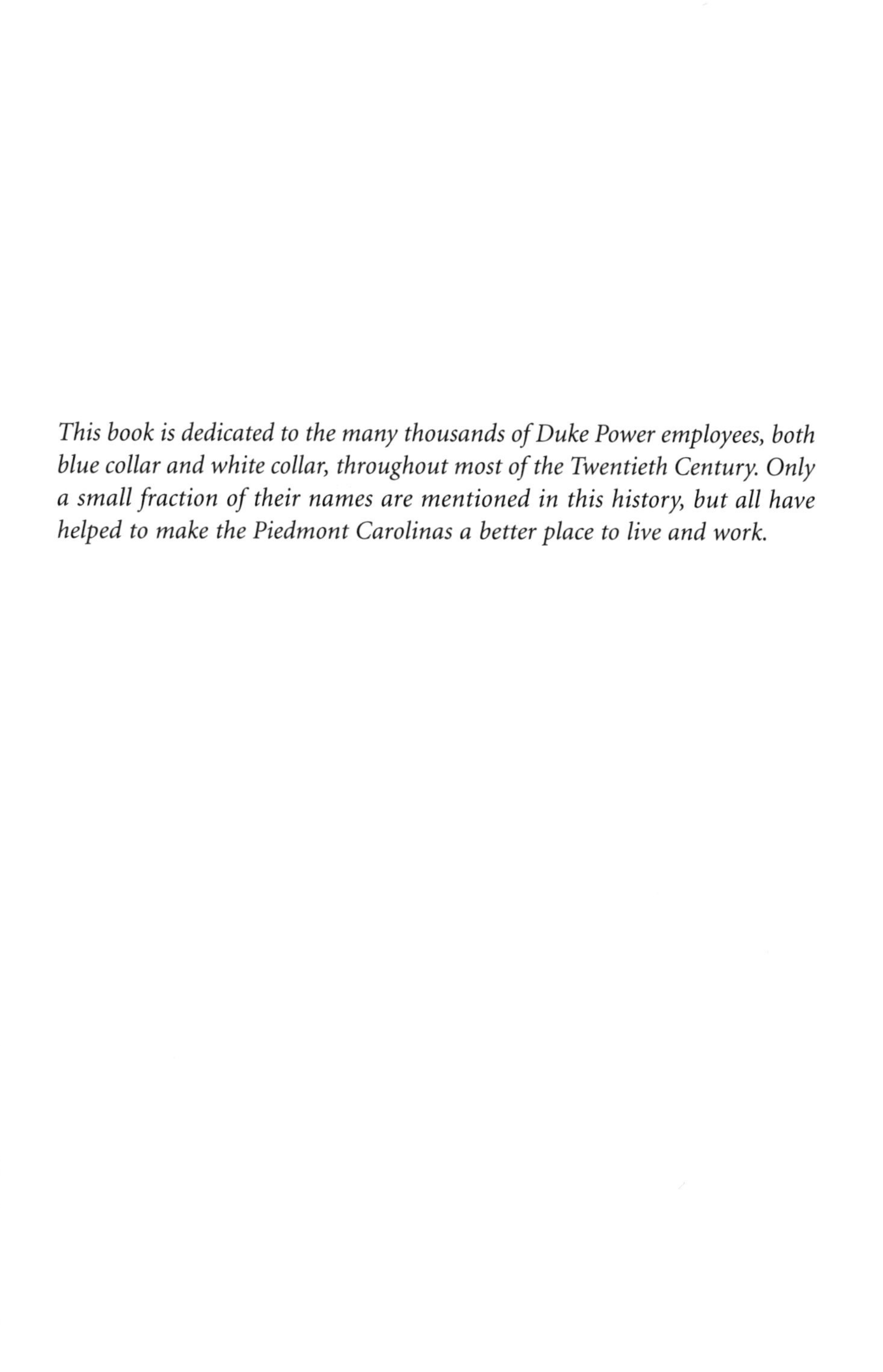

This book is dedicated to the many thousands of Duke Power employees, both blue collar and white collar, throughout most of the Twentieth Century. Only a small fraction of their names are mentioned in this history, but all have helped to make the Piedmont Carolinas a better place to live and work.

Contents

PREFACE

The Duke Power Company's distinctive and proud history, which spans most of the Twentieth century, took a radical turn in June, 1997, when the company merged with one of the nation's largest natural gas companies, PanEnergy of Houston, Texas. While Duke Power continued to be a vital part of the new company, it suddenly became merely a subsidiary—albeit the largest one—of Duke Energy Corporation, as the nationally enlarged enterprise was named.

Beginning with one hydroelectric plant on the Catawba River in 1904, the power company was known first as the Southern Power Company and became the Duke Power Company in 1924. Three men played leading roles in its creation: 1) Dr. W. Gill Wylie, a native South Carolinian who became a distinguished physician in New York, saw great potential in the Catawba's undeveloped water-power sites and also had a most creative idea for linking together a series of hydroelectric plants on the river; 2) William States Lee I, another native South Carolinian and a brilliant engineer educated at the South Carolina Military Academy (later The Citadel), was recruited by Wylie and enthusiastically embraced the idea of linking together multiple generating plants with high-voltage transmission links—if the necessary capital (and lots of it) could be found; and 3) James Buchanan Duke, a native Tar Heel who had made his fortune in tobacco and textiles, saw hydroelectricity as the best means for achieving the industrialization that he hoped could begin to transform the impoverished, overwhelmingly rural and agricultural Piedmont region of the Carolinas. When Wylie and Lee convinced James B. Duke to join with them in 1905, the power company was on its way.

Within a generation, that is prior to J. B. Duke's death in 1925, the company had achieved distinction in several respects. First, most of the early hydroelectric generating plants were built to serve a single city or major factory. On the contrary, the Wylie-Lee-Duke team envisioned from the first an interconnected system of plants and the nation's first comprehensive development of an entire river and its valley. This meant not only a more reliable type of electric service but also a more efficient and economical use of hydro power.

Second, because large amounts of capital were required to build electric utilities, the great majority of those owned by investors ended up under the control of holding companies formed by investment banks and the manufacturers (General Electric and Westinghouse, for example) of expensive turbines and generators. These holding companies, with headquarters in New York or Boston or Chicago, dominated the industry by the 1920s. But the Duke Power Company was a significant exception, for James B. Duke, and to a lesser extent his brother Benjamin N. Duke and his family, supplied most of the capital to launch and sustain the company. Its autonomy meant, among other things, that it was more deeply rooted in its service area—and especially in Charlotte—than was true of most electric utilities.

Third, in 1924 Duke Power, whose engineers had from the first designed the company's plants and dams, began to construct them as well. This was one reason why the company's coal-fired steam plants, which gradually overtook hydro plants in importance during the 1930s and 1940s, long held the national record for fuel efficiency. Moreover, the unique do-it-yourself construction policy carried over into the nuclear era and resulted in significant savings and efficiencies. These, in turn, meant rates to consumers that historically ran about 20 percent lower than national averages.

Fourth, long before such tax-supported, multi-purpose projects as the New Deal's Tennessee Valley Authority, Duke Power's dams and reservoirs, especially but not exclusively on the Catawba, had secondary but significant roles relating to flood control, soil and forest conservation, and recreation. Nature was not generous with lakes in the Piedmont Carolinas, but Duke Power ended up providing eleven lakes on the Catawba alone, and they covered over 70,700 acres.

Fifth and finally, Duke Power became a national leader in its field by pioneering with numerous technologies (such as the use of the nation's first double-circuit 100-kilovolt transmission line). And, as mentioned, Duke's generating plants won recognition over many years as the most efficient in the nation.

The purpose of this book, therefore, is to trace in some detail the evolution—including various trials and tribulations—of a distinctive electric utility through most of the century. The book's thesis is that Duke Power, while by no means the only factor, has been a major player in the economic transformation that has occurred in the Piedmont Carolinas since 1900.

By the 1990s, the Piedmont Carolinas, which had once been a glaringly poor economic backwater, had become something quite different. Many, many problems remained, of course, but few would seriously argue that the problems engendered by poverty and backwardness are to be preferred to those that accompany economic growth and prosperity.

Money magazine in 1994 named North Carolina's Research Triangle the Number 1 place to live in the United States. (Durham, Chapel Hill, and most of the Research Triangle Park itself are served by the Duke Power Company; Raleigh and a small portion of the Park are served by Carolina Power & Light Company). *Entrepreneur Magazine* in 1997 named the Triad (Winston-Salem, Greensboro, and Burlington) the Number 1 best large city in the Southeast for small business (and number 4 in the nation). And the same magazine selected the Charlotte/Gastonia/Rock Hill area as the Number 2 best city for small business in the Southeast. Top or high rankings for all of these Piedmont cities may also be found in other such business-related publications as *Site Selection, Financial World*, and *Inc.*

The economic boom in Duke Power's North Carolina service area continued down the Interstate 85 corridor into South Carolina. There Greenville, Spartanburg, and Anderson could boast high rankings of their own, particularly in the matter of foreign investment. In fact, Greenville and Spartanburg Counties were proud to have the highest per capita foreign investment in the United States.

While the bright economic picture of the Piedmont Carolinas at the close of the Twentieth century resulted from a wide variety of factors—an able and dependable workforce, favorable business climate, excellent transport facilities, and a relatively benign natural climate (thanks in part to air conditioning)—the reliable and comparatively cheap electricity provided by the Duke Power Company has clearly been an important element in the region's economic advancement throughout the century.

~

The way in which I came to write this book is ironic. Putting the finishing touches on my manuscript history of the Duke Endowment in the spring of 1997, I believed that my involvement with Duke-related matters had come to an end—certainly for a while. Consequently, I made a point of stressing in the preface of *Lasting Legacy to the Carolinas: The Duke Endowment, 1924–1994* (Duke University Press, 1998) that the volume completed what I regarded as a trilogy.

The last two chapters of *The Dukes of Durham, 1865–1929* (Duke University Press, 1975) dealt, respectively, with J. B. Duke's establishment of the Duke Endowment in December, 1924, and the beginning of Duke University later in the same month. Accordingly, *The Launching of Duke University, 1924–1949* (Duke University Press, 1993) may be viewed as a sequel to

The Dukes of Durham, with the focus shifting from the family in the latter volume to the new university's first quarter-century in the former one. Then the history of the Endowment continued a story that was barely begun in *The Dukes of Durham.* In a way, therefore, the three books constituted a trilogy, although each could be read independently and each had a different focus and purpose.

This notion of a completed sequence—a neat trio of interrelated books—got somewhat shaken up by an unexpected encounter in April, 1997. William Grigg, a lawyer and then the chief executive officer of Duke Power, invited me to breakfast in Durham and stated that he wished to commission me to write a history of Duke Power. He went on to explain that since not only the name but also the nature and scope of Duke Power were about to undergo a radical change when the merger with PanEnergy became final, he and his associates had decided they wanted to have written the history of the original company.

At first I demurred, explaining that while I had long believed such a history was needed—and had tried in vain to interest several graduate students in the topic over the years—I did not think I was the right person to undertake it. That was so, I thought, principally because I am not a historian of business or technology. In what I thought might clinch my argument, I confessed that I did not totally understand the phenomenon of electricity. Grigg laughingly replied, "Well, I don't either."

That took me by surprise at first, but I quickly surmised that if the head of the company could "wing it" concerning electricity, perhaps I could also. Moreover, Grigg explained that he wanted a book for the general reader rather than for engineers or specialists in economic and business history. He added that he and his associates would be happy to make available all records and check me for accuracy concerning factual matters. All decisions "with respect to the work," however, and "any and all interpretations with respect to events" were to "rest exclusively" with me. (Letter from Wm. Grigg, April 28, 1997).

I agreed to undertake it—and I have been surprised, frankly, by how much I enjoyed the project. In *The Dukes of Durham* (chapter 9) I did deal, in a fairly general way, with the beginning of the power company down to J. B. Duke's death. Then in the history of the Duke Endowment I had again to deal with Duke Power in the 1970s, for it got into some serious financial and other difficulties around 1974–1975. Because Duke Power and the Duke Endowment were so thoroughly entangled with each other, as J. B. Duke had carefully planned for them to be, writing about one necessitated writing about the other. So I was not a total stranger to the history of Duke Power.

True to his word, Bill Grigg has been eminently helpful and supportive. He not only lent me his complete set of the company's *Annual Reports*, but he also arranged an oral history conference in Charlotte in December, 1997—the "Old Rats Meeting" he called it—where an assembled group of some eight or nine top Duke Power executives, mostly retired, reminisced quite colorfully and helpfully for several hours.

Other Duke Power people have also been stalwart supports. Dennis Lawson, the company's archivist, has gone far beyond the call of duty to help; not only did he make several trips from Charlotte to bring portions of the company's archives to Duke University's Perkins Library so that I might use them there, but he has also responded generously and quickly to repeated telephoned requests and inquiries. Likewise, Joe Maher, a senior consultant in public affairs, has been most generous with his help. In addition to procuring for me copies of the histories of several Southeastern utilities, he has read my chapters in draft form with a sharp eye for factual accuracy—and used his copy of the Duke Power *Data Manual* to good effect. Doug Booth, a retired Duke Power executive, also read and made helpful suggestions about several chapters.

At Duke University I am once again grateful to Linda McCurdy and her hospitable, helpful staff in the Special Collections department of Perkins Library for facilitating my research, and my colleague in History at Duke, Bill Holley, has shown a continuing and encouraging interest in the project.

The first two chapters of this book have appeared, in somewhat different form, as a two-part article in the *North Carolina Historical Review*, and I am grateful to its publishers for permission to reprint them.

Andrea Long, assisted at times by Deborah Carver-Thien, did all the word-processing with good cheer; both women even had the grace to express interest in the content, and I heartily thank them for their indispensable assistance.

Anne Oller Durden, my wife, has patiently put up with a lot of conversation (or mini-lectures?) about Duke Power, and I once again gratefully anticipate her help with the tiresome but important chore of making an index.

Robert F. Durden
Professor Emeritus of History
Duke University

Electrifying the Piedmont Carolinas

CHAPTER I

LAUNCHING THE SOUTHERN POWER COMPANY

As the twentieth century started, the South found itself lagging far behind the rest of the nation in most of the positive aspects of modern life. According to virtually all the indices of per capita wealth, literacy, medical care, public schools, and colleges and universities, southern states filled the bottom ranks of the lists. Speaking of the South between 1865 and about 1940, the president of the Southern Historical Association put the matter succinctly in late 1997 when he declared: "In every measure of human progress and welfare, the South lagged far behind the rest of the nation."[1]

The widespread poverty that characterized the region after 1865 was the primary cause for this predicament, which would continue through the 1930s. The two Carolinas were no exception and could take small comfort in being ranked slightly ahead of only Mississippi on many lists. Still predominantly agricultural and rural, the South was the most backward region in a nation that, by 1900, had emerged as the greatest industrial power in the world.[2]

Looking back from the late 1990s, one can clearly discern, however, an area in which the South did not lag. In fact, it was in the vanguard of the movement to make industrial use of electricity. One of the nation's earliest and, at the time, most widely noted ventures in the industrial use of electric power occurred in Columbia, South Carolina. There in April, 1894, a cotton mill began to utilize seventeen electric motors of sixty-five horsepower each, and the alternating current that ran them was furnished by a hydro-station located on a river about a thousand feet from the mill. At $15 per horsepower, the Columbia mill was reported to be utilizing the cheapest power in the nation.[3]

Moreover, an investor-owned electric utility began to take shape in the South in 1904–1905 that would not only play a major role in the industrial transformation of the Piedmont Carolinas but would also become a national leader in the electric utility field. Starting off as the Southern Power

Company and becoming the Duke Power Company in 1924, it would be distinctive and different from most other investor-owned utilities in several ways.

First, from the beginning the men who built the company envisioned not scattered hydroelectric generating plants, each serving a single city or major factory, but rather a series of generating plants, interconnected by high-voltage transmission lines, that would form a single system. This would be the nation's first comprehensive development of an entire river and its valley for the generation of power and other beneficent purposes. Not only would this be a way of providing more reliable electric service, but it would also be a more efficient and economical use of hydro power. Furthermore, the company early established power linkages with other utility companies and gained national recognition for creating a pioneering "system of systems."

Second, because large amounts of capital were required to build electric utilities, the great majority of those that were investor owned ended up in hock to the manufacturers of expensive generators and turbines or to investment banks. By the 1920s, utility holding companies, with their headquarters in New York or Boston or Chicago, dominated the privately owned electric utility industry.[4]

The Duke Power Company, however, was different. James B. Duke, and to a lesser extent his older brother Benjamin N. Duke, supplied most of the capital required to launch the company. From the beginning, it enjoyed—and kept—an autonomy that was highly unusual among the major, investor-owned electric utilities. This autonomy had profound implications for the manner in which the company operated and for its financial health. The Duke Power Company would prove to be more deeply rooted in its service area (and especially in Charlotte) than was the case with most utility companies. And when holding-company empires such as that of Samuel Insull came crashing disastrously down in the early 1930s, the Duke Power Company escaped unscathed.

Third, in 1924 Duke Power, whose engineers had from the beginning designed the company's plants and dams, began to construct them as well. This unique do-it-yourself policy would continue into the era of atomic energy in the latter decades of the century and would result in outstanding efficiencies and significant savings. These, in turn, meant rates to consumers that were well below national averages.

Fourth, long before the Federal government touted the multi-purpose nature of such tax-supported New Deal projects as the Tennessee Valley Authority in the 1930s, Duke Power built a series of dams and reservoirs, especially on the Catawba River, that, in addition to their primary role in power generation, had secondary roles relating to flood control, conservation of soil and forests, and the provision of recreational opportunities. In a region where natural lakes were sadly lacking, Duke Power ultimately provided eleven lakes on the Catawba River alone; they covered a total of about 70,715 acres, with Lake Norman near Charlotte (about 32,500 acres) being the largest body of fresh water in North Carolina.

Fifth and finally, Duke Power became a national leader in its field by pioneering with numerous technologies and having its generating plants recognized by trade journals and industry associations as the most efficient in the nation over a period of many years. Some of these plants also won prestigious national awards as outstanding engineering projects.

In addition to James B. Duke, several others played key roles in starting the company, and in order to show this, perhaps the best place to begin is in the then small town of Anderson, South Carolina, in the 1890s. There, in the upcountry near the Savannah River, a young engineer named William Church Whitner began to be active on the frontiers of industrial electricity. Born in Anderson in 1864, Whitner graduated from the University of South Carolina in 1885. He immediately returned to the university, however, for an additional degree in civil engineering. After working as a railroad engineer, young Whitner agreed to undertake in Anderson the construction of a system of waterworks as well a steam-powered electric-light plant.

Following Thomas A. Edison's invention of the electric light bulb late in 1879, electricity and particularly electric lighting captured the imagination of much of the American public. Cities and towns across the nation clamored for this latest symbol of progress and modernity. Electric streetcars were not far behind, and in 1883 Montgomery, Alabama, and Richmond, Virginia, boasted of the first electric tramways in the United States. Given the rudimentary nature of much of the machinery and technology involved in these early ventures into electrification, one should not be surprised that electric service was both erratic and expensive. An historian of one utility notes that in the early days, electric lighting service was available in many communities only during limited hours. Many of the street-lighting companies operated on a "moonlight basis": when the moon grew bright enough to cast a shadow, off went the street lights.[5]

In Anderson, Whitner finished both of his plants on schedule, and the Andersonians took great pride in their 750 incandescent lamps. Whitner, however, was not altogether satisfied. Coal, not available in the area, was expensive and the two plants he had constructed were not proving to be profitable to the owners. He therefore began to study and search for a less expensive method of producing electricity than the use of coal-fired steam. "In 1891," Whitner later wrote, "I became convinced, on account of experiments that had been made in Europe [Germany] in the development and transmission of electric power from waterfalls, that it would soon be possible to utilize some of the fine waterpowers in the vicinity of Anderson."[6]

Whitner, not content with studying his technical journals and papers, traveled to New York to obtain first-hand information from an important participant in the fast-breaking development of industrial electricity. Nikola Tesla was a Croatian immigrant who had worked for and then been fired by Edison. Despite all of his pioneering creativity, Edison bet on the wrong horse when he stuck stubbornly with his direct current (DC) electric system. Direct current could not be economically transmitted for any distance, and the rival alternating current (AC) system was winning out in the much-discussed "Battle of the Systems" of the late 1880s. By inventing the alternating-current electric motor in 1888, Nikola Tesla helped in the eventual defeat of his former employer's direct current system.[7]

Returning to Anderson after the meeting with Tesla and a study of available electrical machinery in various northern cities, Whitner decided to attempt to build a hydroelectric plant at a waterpower site known as High Shoals on the Rocky River six miles from Anderson. With so many scoffing at his idea of transmitting electric power over six miles of bare wire, Whitner had great difficulty raising the necessary capital. He persevered, however, and finally persuaded the owners of the Anderson Water, Light, and Power Company to provide at least enough capital for a small, experimental hydro-station. "Electrical machinery for such work was in the experimental state," Whitner later explained, "and it was quite a difficult problem to decide what to use, but being convinced that high voltage should be employed to as great an extent as possible, I purchased from the Stanley Electric Manufacturing Company, of Pittsfield, Mass., [a] 120 K. W. [kilowatt] alternating current generator wound with all necessary apparatus."[8]

Whitner's experimental plant began operating in May, 1895. In the following month the *Anderson Intelligencer* proudly announced that it was printed on a press run by electricity. "Can any of our contemporaries in the State or in the South say as much?," the editor queried. "We feel very proud," he continued, "to be able to boast of having the pioneer company in the State, or indeed in the South, as we are informed, to demonstrate the practicability of electric power by what is known as the long distance transmission system; and we hope that our contemporaries will take due notice of the fact and give to us our due meed of praise."[9]

The boastful editor was not the only Andersonian to be pleased with Whitner's achievement. The owners of the Anderson Water, Light and Power Company were sufficiently impressed to authorize an increase in the capital stock and bonded indebtedness of the company so that Whitner could undertake a much more ambitious hydroelectric project. This time he proposed to build a generating plant at Portman Shoals on the Seneca River, ten miles from Anderson. (The Seneca joins with the Little River to form the Savannah River.) Having first made a tentative contract to supply power to the Anderson Cotton Mills, Whitner now planned for a 4,000 horsepower development. His earlier experiment had convinced him that high voltage generators should be used in order to avoid the necessity of expensive step-up transformers. Whitner hoped to use generators wound for 10,000 volts if he could obtain them. The difficulty was, however, that no such generators had yet been built for commercial work, and the large manufacturers refused to build them. Whitner finally persuaded the Stanley company, with which he had dealt earlier, to build two generators wound for 10,000 volts and having a capacity of 900 horsepower each. "These were the first alternating current 10,000-volt generators ever built for commercial work," he later noted, "and, of course, they were regarded as more or less of an experiment."[10]

After construction began at Portman Shoals in 1896, Whitner hired another young engineer as his assistant. This was William States Lee, whose name was destined to loom large in the history of Duke Power. Born in Lancaster, South Carolina, in 1872, Lee grew up on a farm near Anderson after his family moved there. Although young Lee hoped to go on to college after completing his work in the Anderson public schools, he was one of nine children in a poor family, and college seemed out of the question. Unwilling to give up, Lee entered a competitive examination for a scholarship at the South Carolina Military Academy (later The Citadel) and won it.

The rail fare of $7.65 from Anderson to Charleston posed a problem, Lee later recalled, but his supportive and determined mother helped him scrape together the fare. When he got to Charleston, Lee found that his "clothes were not in keeping with things there," but he managed to gain a degree in engineering "without worrying much about clothes." Moreover, when he graduated, he had two new suits, the price of which he had saved "from the allowance the State had made for... board, books, and tuition."[11]

Since Lee's scholarship obligated him to teach in the public schools for two years, he fulfilled that requirement by teaching during the summers and found his first engineering work with a railroad company. The opportunity to work with Whitner on one of the earliest hydroelectric projects in the South soon arose, however, and Lee found himself on the cutting edge of hydropower development.

Exactly how much Lee learned from Whitner, who was eight years older than Lee and more experienced, can only be surmised, but it may have been a good bit. At any rate, the two men must have worked well together, for the plant at Portman Shoals went into successful operation in November 1897, clearly demonstrating the feasibility of high voltage generators. "The success of the plant (which was the first long distance hydro-electric transmission plant built in the South)," Whitner later noted, "brought the undeveloped water powers of the State [and the Southeast in general] prominently to the notice of the public, and many [water-power sites] were surveyed and bought by investors."[12]

Prominent among those investors who had been awakened to the new potential of water-power sites were two wealthy Tar Heel entrepreneurs, James B. Duke and his older brother Benjamin N. Duke. As will be discussed below, their engineers and agents were in the process of acquiring certain highly promising sites in the Piedmont Carolinas.

As for Whitner, he moved on from his success at Portman Shoals to design and build a major hydroelectric project on the Chattahoochee River at Columbus, Georgia. Once again he employed W. S. Lee as his assistant engineer, so that both men added to their laurels as pioneers in the new field of hydroelectric engineering.

According to a distinguished historian of technology, the half century from 1880 to 1930 constituted the formative years of the history of electric supply systems. Concerning the achievement of those years, he declares: "Of the great construction projects of the last century, none has been more impressive in its technical, economic, and scientific aspects, none has been more influential in its social effects, and none has engaged more thoroughly our constructive instincts and capabilities than the electric power system."[13] From a global perspective, William C. Whitner and William States Lee were not, of course, major players in the achievement that the historian hails. From a regional perspective, however, they were, and their roles as pioneer hydroelectric engineers in the South were important. While Whitner left Lee to manage the large project at Columbus, Georgia, it was Whitner who came up with an idea that would lead directly to the beginning of the Southern Power Company. The Catawba River, rising in the mountains of North Carolina and flowing down through the Piedmont into South Carolina (where it became the Wateree River), presented a more challenging—but potentially more promising—project than he had so far tackled. After surveying the various parts of the Catawba south of Charlotte, where there were a number of rocky declivities or falls, Whitner decided that an old watermill site at India Hook Shoals was what he wanted.[14] It was six miles north of Rock Hill, South Carolina, and sixteen miles south of Charlotte. The proximity to Charlotte was, Whitner recalled, the main attraction, for it was already the most important city in the area. Securing options on most of the needed property around India Hook Shoals, Whitner then tackled the problem of finding the necessary capital for his envisioned dam and hydroelectric plant. For that, he travelled to New York City to talk with someone who had earlier purchased stock in and was much involved in the Anderson Water, Power and Light Company and who owned the Portman Shoals site—Dr. W. Gill Wylie.

Wylie's important role in the beginning of what would become the Duke Power Company grew obscured for some years after his death in 1923. Gradually, however, the historical record became clearer, and from around 1960 onward Wylie's contribution has been fully acknowledged. In belated recognition of that contribution, the Catawba Lake (or reservoir) and Catawba Hydro-station were renamed in 1960 as Lake Wylie and Wylie Station.

Born near the Catawba River in Chester, South Carolina, in 1848, Wylie at age sixteen entered the Confederate army late in the Civil War. As a young lieutenant, he led a company of other young men who participated in the failed effort to halt General William T. Sherman as he led his conquering Federal army through South Carolina early in 1865. Despite the havoc caused by the war, young Wylie persevered in his ambition to become a medical doctor, like his father. Taking all the available scientific subjects, including civil engineering, at South Carolina College (later the University of South Carolina), Wylie graduated in 1868 and headed directly for New York City. There he entered Bellevue Medical College and received his medical degree in 1871. After a year on the medical staff of Bellevue Hospital, he travelled to England and the continent with a special interest in studying surgical techniques, hospital construction, and the nurses' training schools in Britain begun earlier by Florence Nightingale.

Returning to New York in 1873, Wylie established a private practice; helped to establish the Bellevue School of Nursing (the first in the United States to be based on the Nightingale plan); and in 1882 began a long association as the visiting gynecologist at Bellevue Hospital and as a professor of gynecology at the Polyclinic School of Medicine, where he also lectured on abdominal surgery for more than twenty years.

W. Gill Wylie, in short, became an eminent and well-to-do medical leader in New York. Just how or why he became interested in hydroelectric developments in the Carolinas is not known. Many years later, his son conjectured that Wylie's friendship with Charles Steinmetz, a leading electrical researcher and theoretician who joined the new General Electrical Company in 1892, had stimulated an interest in the new field of industrial electricity. The son also recalled that his father had known Thomas A. Edison. At any rate, Wylie had grown up close to the Catawba River which, he later said, "with the exception of the Penobscot River [in Maine], is the best river east of the Rockies for power development."[15]

Wylie, like J. B. Duke and others, may have envisioned hydroelectric development as a way whereby his native region could at least begin to escape from agrarian poverty by turning to industrialization. He, like other educated Americans in the early 1890s, no doubt watched with great interest as the "Battle of the Systems" raged concerning the large hydroelectric development that was underway at Niagara Falls. In 1893 J. P. Morgan and his associates chose George Westinghouse's alternating current system, rather than Edison's direct current system, for the Niagara Falls project; and when the Niagara generating station was completed in 1895, a veritable parade of industries moved in to take advantage of the new and cheap electric power.

Hydro-electric power, in short, was much in the news in the 1890s, and Wylie's thoughts on the subject turned to his native state. Wylie later noted that he had consulted his friend Steinmetz concerning the best electrical machinery and transmission systems for the Carolina projects.[16] At any rate, W. Gill Wylie and his younger brother, Robert H. Wylie, who was also a gynecologist in New York, had earlier made a substantial investment in the Anderson Water, Power and Light Company and played an important role in the building of the hydroelectric plant at Portman Shoals that had attracted such widespread attention. Accordingly when Whitner arrived in New York with the idea about a new project on the Catawba River at India Hook Shoals, the Wylie brothers were quick to buy into the plan.

In 1900 the Wylies and Whitner organized the Catawba Power Company, with W. Gill Wylie as president and Whitner as general manager and chief engineer. The construction contract for the Catawba dam and powerhouse was soon signed but, unfortunately, "many great floods" on the river during 1901 caused the work to proceed slowly. W. Gill Wylie later recalled that during that period he studied the whole matter further and made some changes in the original plans. Whether these changes triggered the falling out of Wylie and Whitner, after the delays and frustration of 1901, is not known. What is known is that Whitner resigned from his position in 1902, sold his interest in the Catawba project to the Wylies, and went on to have a productive and important career in Richmond, Virginia, as chief engineer for the Virginia Railway and Power Company. Casting about for a replacement, Wylie first tried to hire a government engineer from Columbia, South Carolina; he turned down the offer but suggested that Wylie should employ William States Lee, Jr., who had accredited himself well on the project at Columbus, Georgia. Lee promptly accepted Wylie's offer of $4,000 a year, which was a substantial salary in 1903. Referring to his initial conversation with Lee, Wylie later made an important statement. Since Wylie was speaking at the 7th annual banquet for employees of the Southern Power Company in late December 1912, Lee was probably in the audience and heard Wylie make this claim:

> I explained to Mr. Lee at that time [March 1903] that if he succeeded with the Catawba plant, his reputation as an engineer would be made, and I explained to him at that time the scheme I had for building dams all along the Catawba so that we could utilize a large part of the 700-foot fall which occurred through its length of 130 miles from Camden, S.C., [up] to Hickory, North Carolina.[17]

Since the idea of multiple, interconnected hydroelectric plants along the Catawba River was destined to become a centrally important and distinctive feature of the Duke Power system, Wylie's claim to authorship of the plan must be carefully noted. Certainly, however, no one accepted the vision more quickly or with more enthusiasm than young William States Lee.

In an interview that Lee gave a decade or so after Wylie's 1912 speech, Lee explained to the journalist about some of his early ideas in relation to Wylie's vision. Lee first pointed out that a few miles below the original Catawba plant were the Great Falls, with a "head" or fall of 160 feet; then below that part of the river was Wateree, with a fall of 72 feet. To build power plants at these falls and tie them in with the Catawba plant at India Hook Shoals would, in effect, be laying the foundation of great superpower system. "I didn't know where the money was coming from to finance this scheme of mine [and Wylie's]," Lee declared, "but I didn't trifle or neglect the details while waiting for something to happen. You never know when opportunity is going to knock at your door. . . ." Lee noted that he "worked just as hard on that super-power idea as if I had unlimited capital at my command." He said he surveyed every yard of the Catawba River and had his plans all worked out in minute detail before knowing how the scheme was going to be financed. It even occurred to him, Lee confessed, that nothing might ever come of the plan, but he was utterly convinced of its practicality. "It was something that just *had* to be done," Lee believed, "and it was a job for some man who would give his heart and soul [and a large chunk of his money] to it."[18]

That rich man willing to commit to and underwrite the Wylie-Lee vision of a "super-power system" would, of course, eventually turn up. Before dealing with that, however, developments at the Whitner-Wylie power plant at Portman Shoals pointed up certain hazards in the early days of industrial electrification when a manufacturing company had to depend on a single, stand-alone power plant. Or, to put the matter another way, the greater reliability of service (not to mention economies and efficiency) that would derive from a series of interconnected power plants was well pointed up by certain events at the original Portman Shoals plant.

The minutes of the board of directors of the Anderson Cotton Mills on December 31, 1901, noted that two days earlier about a third of the dam at Portman Shoals had been washed away by high water caused by heavy rains. This accident—and consequent suspension of electric service—forced the temporary shutting down of the Anderson Cotton Mills, but power was expected to be restored within four months.[19]

As those who managed the early Duke Power plants would gradually learn, however, hydropower trouble could result as much from droughts as from floods. On October 10, 1905, the board of directors of the Anderson Cotton Mills submitted their annual report to stockholders and expressed "regret that circumstances over which we had no control prevented our making the amount of profit for the year we would have made under other conditions." There had been unprecedented low water in streams (and particularly on the Seneca River) throughout the area during the summer and fall of the previous year. The No. 1 mill had been stopped (by power outages) 41 times during the year, which caused a direct loss of 22 days' time. The No. 2 mill had 82 stoppages, which lost almost 26 days. The indirect loss was more than the actual time stopped, the director continued, "as our help left because of our not being able to give them steady employment and after the normal condition of the water was restored . . . a long while was required to get back enough help to even furnish our mill a scant supply."[20]

Since W. Gill Wylie was so much involved with the plant at Portman Shoals, experiences such as those described above may well have been a major inspiration for his vision of a series of interconnected hydroelectric power plants. And although Wylie had recruited W. S. Lee in March, 1903—and found an ardent and hard-working sharer of the larger vision—the Catawba plant at India Hook Shoals was far from complete. Dreams of plants at Great Falls and Wateree were premature, to say the least, when the original Catawba plant had yet to be finished.

In 1903, Wylie collided head on with a basic fact about the electric power industry: it was a cash gobbler or, as economists would prefer, the industry was highly capital intensive. Tobacco growing, for example, was—and is—highly labor intensive; but money, not labor, was the great need in the electric power industry. In Wylie's case, the dam at India Hook was nearing completion in early 1903, but he realized that he needed an additional $250,000 to build and equip the powerhouse. The Catawba Power Company had an authorized bond issue of $500,000, of which $400,000 had been sold. Wylie wanted authorization for an additional bond issue, but the subscribers to the previous issue would not consent to anything beyond the sale of the additional $100,000 in bonds that had already been authorized.

Thus, by February 1903, Wylie recalled, it had become evident that some other financial arrangement had to be made. "At that time," Wylie declared, "I believe I was the only person alive who had any faith in the final completion of that development."[21]

Finally, with the help of investment bankers in Philadelphia, who sent their own engineer down to check on the Catawba project, Wylie secured the financial assistance he needed. He and his brother had put about $350,000 of their own money into the project, and Wylie believed that helped convince the bankers of his and his brother's earnestness. W. S. Lee proceeded energetically to complete the Whitner-Wylie plans for the project, and by March 1904, the first hydroelectric plant on the Catawba River was completed. It had a capacity of 6,600 kilowatts.

Shortly before dawn on March 30, 1904, the control or head gates in the dam were opened, and tons of water went crashing downward through the wicket gates on to the spiraled blades of a water turbine, or as it was then still called, a waterwheel. As the blades turned, ropes and pulleys transmitted the water's energy to an electric generator, wherein wires wrapped around a magnet produced the "miracle" of electricity.

The first and, for a brief time, only customer for the new power was the Victoria Cotton Mills at nearby Rock Hill, South Carolina. The power was delivered to the factory over a 11,500-volt line strung on wooden poles, and it energized a 300-horsepower motor for operating the entire cotton mill. An entry in the Catawba plant log made on that first day has long amused Duke Power people, for it points up the stark simplicity then of an industry that would gradually become infinitely more complex and sophisticated: "This being the only load [to the Victoria Cotton Mills], started one generator up at 6:00 A. M.—shut it down at 6:30 P. M., locked the powerhouse door and went to bed."[22]

Lee, in his report to Wylie and the stockholders of the Catawba Power Company in June 1904, noted that the power plant had given "good service" to the Victoria Cotton Mills, and "they are well pleased, and the use of this Power shows a great saving over Steam." Already power lines had been extended to other plants in Rock Hill, and about two miles of the Charlotte pole line had been completed. Lee expected to have poles erected all the way to Charlotte and ready for wire by July 1, 1904. "There is no doubt but that we will do a good lighting and small Power business just as soon as we are able to deliver the current," Lee declared.[23]

In Lee's report a year later, he was even more optimistic about power sales. The company's lighting and small power service had both been "very satisfying as well as profitable." The small motor service had begun with quite tiny units from a fraction of a horsepower up to twenty horsepower." Since our customers have seen the advantage of the electrical drive," Lee added, "we are now getting a great many customers who will use from 50 to 100 H. P." Lee believed that the public in general was becoming "much more favorably impressed with the advantage of electric power" and better informed about the high cost of steam power.[24]

Slowly and painstakingly, then, the cornerstone of what would become a vast utility system was being built. Even by the standards of the late twentieth century, the building of a power generation and transmission system is a gigantic undertaking. Given the primitive technology and equipment of the early 1900s, however, it must have been a daunting task. Often miles from the nearest town, construction crews lived in tent camps as they carved the narrow corridors through which power would flow to the cities and towns of the Piedmont. Using mules, oxen, and their own musclepower, they erected the poles of cypress and chestnut and strung the copper wire to transmit the electricity.

W. S. Lee later reminisced about some of his early experiences with the power company. According to him, the men stringing a transmission line were bungling the job and holding up the work. "I took over the job myself," he explained, "and strung the line in about half the time people thought it would take." From that incident derived the legend that Lee started as a linesman and worked up to the job of chief engineer. On the contrary, Lee noted, "I started as engineer and worked *down* to linesman! It has been my idea never to let anything hold up the work. If the job is suffering and you haven't a man to put it across, do the job yourself." Expatiating further on his philosophy of work and leadership, Lee added: "I've learned this about men. It isn't preaching that inspires one's helpers so much as doing. Almost anybody can preach and get some sort of hearing.

But red-blooded workmen take their hats off to the man who can show them." Men gain their faith in a leader, Lee concluded, when he shows rather than tells him what sort of person he is. "I never had to tell anybody that I was boss of the works."[25]

With the Catawba Power Company off to such a promising start, Lee and the Wylies had every reason to be proud. They were about to enjoy an unforeseen stroke of good luck, however, when the Dukes got into the act. Contrary to an oft-told tale, the involvement of the Dukes in hydroelectric developments in the Carolinas was not altogether the result of Dr. Gill Wylie's being asked to treat J. B. Duke's inflamed foot. Cautious entrepreneurs, J. B. and B. N. Duke never invested their money capriciously or on the spur of the moment. Using capital made first in the tobacco business, the Duke brothers went into textile manufacturing in Durham and elsewhere in North Carolina in the early 1890s. They selected as the general manager of their textile operation an unusually able and dependable person, William A. Erwin. Consequently, the Dukes and Erwin were well aware of the significant developments that were underway in the 1890s concerning hydropower and industrial electrification. In addition to hiring their own engineer to look over possible waterpower sites and to advise in the new field, the Dukes had the faithful assistance of Erwin.

When the engineer occasionally revealed certain information to outside parties, Erwin complained, for he had the true businessman's penchant for secrecy and quiet operation. "I am of the opinion," Erwin wrote Ben Duke, "that 'much learning' makes hydraulic engineers and preachers mad: most of them needed guardians...." Yet Erwin, liking the engineer and believing him to be competent, hoped that he would learn to be "more politic." Erwin added that he was "making a special effort quietly to learn the status of the Catawba River [Great] Falls, which is unquestionably the biggest thing in the South...." A little over a year later, in 1901, he had managed to acquire the coveted site for a little less than $42,000.[26]

After several years in which Erwin investigated numerous power sites in the Carolinas, he concluded, quite prophetically, that the "main trouble with all these powers" was the erratic flooding and drying up of rivers in a region where periods of heavy rainfall often alternated with prolonged droughts. The one exception, in Erwin's view, was "the Great Falls of the Catawba, where the river never rises above eleven feet, and [which] would give practically an invariable steady power." It was, Erwin asserted, "the greatest property... in the United States on this account," as well as because of the "low cost of development it would require...."[27] In 1899 the Dukes incorporated the American Development Company to secure and hold their land and water rights, primarily along the Catawba River. Dr. W. Gill Wylie tried, in vain, to interest the Dukes in the initial project on the Catawba at India Hook Shoals. Ben Duke was a patient of Wylie's, and after thanking Ben Duke for a "generous addition" to a bill for a successful appendectomy, Wylie again in October 1899, appealed to the Dukes for assistance. He explained that he and his brother had "bought now and have the deeds to most of the lands necessary to give us a fall of full 18 ft.... We will have to raise about $20,000 sure to complete the purchase and if you don't come in with us or your [American Development] company buy us out we will have to borrow the cash and go it alone... Bob [Wylie] and I would rather go in with you and your brother as individuals than sell out to any one else. From our experience at Anderson I am sure we have a good thing but it is too much for us to carry out alone...."[28]

Although Ben Duke assisted Wylie, the Duke brothers had tobacco and textiles at the center of their attention around the turn of the century and simply were not ready for extensive involvement in the new hydroelectric power industry. Not one to be easily shaken off, however, Gill Wylie persisted. When he ran into his financial emergency with the unfinished Catawba plant in 1902, he again turned to Ben Duke. Admitting that he was embarrassed in light of all that Ben Duke had already done for him, Wylie explained that "the Charleston men who have taken up the financing of the Catawba Power Company urged that your name would be of great help to us in S. C. where we desire to have a good portion of our bonds and stock held...." After noting that he and his brother had already put about $300,000 into the project, Wylie concluded: "This kind of letter is new to me, and I have such a warm affection for your kind and gentle self that I really have for so few men that I have ever met that I am afraid you may class me with many others that have approached you." Wylie did not appeal in vain, for Ben Duke subscribed for $25,000 worth of bonds of the Catawba Power Company.[29]

Not B. N. Duke, however, but J. B. Duke was the brother with whom Wylie needed to establish a relationship. This was because, while B. N. Duke was the older brother and took the lead in the family's many charitable activities, it was J. B. Duke who was the true entrepreneurial genius of the family. This had become apparent as far back as the early and mid-1880s in the tobacco industry. The two brothers were remarkably close to each other, and to their long-widowed father, but even Ben Duke would have admitted that he yielded to "Brother Buck" when it came to large decisions concerning business and investment.

Where business matters were concerned, not only could J. B. Duke think faster and more clearly than most people, but he also had a special knack for "thinking big." Furthermore, he had no hobbies or diversions, other than having extensive landscaping done at his estate, then known as Duke's Farm, near Somerville, New Jersey. Tremendously enjoying his work, whether building up an empire in tobacco or an electric utility in the Carolinas, he was what a later generation would call a happy "workaholic." The poet Robert Louis Stevenson once had this to say about such a person as J. B. Duke: "If a man loves the labor of any trade, apart from any question of success or fame, the Gods have called him."

J. B. Duke was clearly one whom "the Gods" had called, and finally in the fall of 1904 and spring of 1905, Dr. Gill Wylie got his chance to do some talking with him. J. B. Duke was never an easy person to pin down for an appointment and wrote as few letters as possible, but he developed an inflammation in his foot and called in Dr. Wylie for treatment. By this time, the Dukes had been marginally involved in Carolina hydropower developments for more than five years, so Wylie was not speaking to a total novice about his vision for the development of the Catawba and the great ability of his chief engineer, William S. Lee, Jr. After hearing Wylie talk about his vision, J. B. Duke invited him to have Lee come up to New York with the surveys, preliminary plans, and other data that he had been so carefully assembling.

Perhaps another prominent businessman's personal assessment of J. B. Duke, at a later stage in his life, might be relevant here, for there is no reason to think that Duke's basic manner and personality changed over the years. This later observer noted that J. B. Duke, in addition to being alert and possessed of a sharp mind, was a "soft-spoken" person, a "courteous Southern gentleman" who seemed "very friendly and very keen."[30]

There is every reason to believe, therefore, that J. B. Duke gave Lee a courteous hearing when he came with Wylie for the interview. Lee later recalled that Duke, after posing sharp questions and examining Lee's plans and diagrams, asked what the estimated cost might be. "I told him about $8,000,000," Lee noted. "I thought that was about the biggest amount I had every heard of, but it seemed to attract him."[31]

Lee quickly realized that hydroelectric power for industrial uses was the aspect that most interested Duke. Since long-distance transmission of electricity at high voltage was still in the early stages, many financiers were fearful of the risks involved. Lee discovered that Duke, because of his keen interest in promoting the industrialization of the Piedmont Carolinas, was willing to take the risk if practical plans were prepared. On that point Lee was ready, for he had prepared a map showing transmission lines that tied together the existing Catawba plant and projected plants at Great Falls and at Mountain Island, which was on the Catawba to the northwest of Charlotte and more than fifty miles north of Great Falls. Lee, giving practical application to Wylie's original vision, aimed at continuity of service through linking the various plants together. There was also another site, Wateree, down the river some eighteen miles below Great Falls, which Lee envisioned as part of the system.

J. B. Duke asked Lee and Wylie many sharp questions, which he always preferred to reading documents. "I know that in many cases he never had studied or heard of the things brought up," Lee recalled, but "readily grasping the idea, his mind passed on to the next step." Disdainful of committee meetings or conferences involving what Duke regarded as too much talk—"town meetings" he termed them—he quickly made up his mind and verbally issued many important instructions. "He had a wonderful power of making decisions," according to Lee. Sometimes they "seemed to be almost off-hand," but "they were as accurate as they were swift." Lee concluded his admiring appraisal thusly: "Generally, he had gone into the matter thoroughly, had the points fixed in his mind and was sure of his ground. He merely thought faster, more accurately, and grasped the points of a situation more quickly than most men. And, once he had decided, he acted promptly."[32]

After J. B. Duke's interviews with Wylie and Lee and after several years of cautious preliminary investigation, the Duke brothers were ready by the spring of 1905 to get started in the electric utility industry. The Southern Power Company was incorporated in New Jersey on June 22, 1905, with an authorized capital stock of $7,500,000. By March, 1906, $6,000,000 worth of stock had been issued in exchange for $1,097,794 in cash (mostly from J. B. and B. N. Duke); the entire $850,000 worth of the capital stock of the Wylie's Catawba Power Company; a demand note from that company to the Southern Power Company for $118,579; and property, real estate and water rights valued at $3,933,607 (mostly paid for by the Dukes).[33]

The new Southern Power Company located its principal office in Charlotte and had as its officers W. Gill Wylie as president, B. N. Duke as vice president, and W. S. Lee as chief engineer and general manager. The company lost no time in launching an extensive construction program that would not stop for any significant duration until the Great Depression of the early 1930s.

The Great Falls of the Catawba that had appeared so splendidly promising for development in the eyes of W. A. Erwin had long attracted attention. What excited the pioneers of hyroelectricity had, in the early nineteenth century, dismayed other ambitious men, for the falls in so many southern rivers impeded navigation. Consequently, in the Carolinas, as in many other states, there was a great and ultimately unsuccessful push to build canals and locks that would overcome the navigational problems posed by the falls. By the 1840s most of these efforts had failed, and the railroad became increasingly important in transportation, lessening the need for canals.

Not only were there the relics of the abandoned canal at Great Falls, but the Federal government had built a short-lived arsenal there in the early 1800s during President Thomas Jefferson's administration. When the last soldiers left the arsenal (Fort Dearborn) in 1817, they left behind a massive building made of huge granite blocks. By 1905, when the power company people arrived on the scene, they found large trees growing through the space where the roof of the granite building had long since disappeared. Power company engineers restored the roof, reclaimed the building, and used it as a club house.[34]

Aside from the derelict monuments of past eras, the Great Falls themselves yet retained the grandeur that had, through the years, thrilled those who saw them. In the early nineteenth century, one observer left a classic, almost musical description:

> ...Hills confine the descending stream as it approaches them [the Great Falls]; when advancing nearer, it is further narrowed, on both sides, by high rocks, piled up like walls. The Catawba River from a width of 180 yards, is now straightened into a channel about one-third of that extent, and from this confinement is forced down into the narrowest part of the river, called the gulf. Thus, pent up on all sides, on it rushes over large masses of stone, and is precipitated down the falls. Its troubled waters are dashed from rock to rock, and present a sheet of foam, from shore to shore, nor do they abate their impetuosity until after they have been precipitated over 20 falls, to depth of very little short of 150 feet. Below Rocky Mount the agitated waters, after being expanded into a channel of 318 yards wide, begin to subside, but are not composed. A considerable time elapses before they regain their former tranquility.[35]

The Southern Power Company's first foray into plant construction took place, therefore, at a spot of great natural beauty and haunted by history. But in August 1905, as construction began, the men at Great Falls were not concerned with the past but focused on the future. In order to move materials and equipment to the plant site, the company first had to build twelve miles of railroad. Then with a labor force that at times numbered as many as 1,000 men, and averaged nearly 600, they frequently worked day and night to construct the dam and powerhouse.[36]

By April 1907, the 24,000-kilowatt plant, with a total original cost of a bit more than $1,612,000, was ready to go into service. A 44,000-volt transmission line delivered power first to Charlotte and then later the lines were extended to Gastonia, Shelby, and other towns in North Carolina. The total surface area of the reservoir behind the Great Falls dam was 400 acres at full pond, and the property holdings for the development totalled nearly 9,500 acres. Since much of the land was unsuitable for agricultural leases, the company planted 750 pecan trees on a portion of it, hoping that in time they would bring in revenue as they also helped with soil conservation and reforestation.

A reporter for the *Charlotte Observer* visited the station in the spring of 1908 and thought it a "work of massive beauty." The engineering involved, according to the reporter, had been "well-nigh perfect, being so arranged as to turn every drop of water that the river affords into the canal and delivering same with an even flow to the eight massive turbine water wheels. . . ." At the same time, "mechanical arrangements have been perfected through which, in case of high water or flood, the excess water can be forced to escape over a large spillway, and thence to flow around the power house—an ideal arrangement this." The newspaperman may have been a bit over-optimistic when he added that the "head-gates and wheel pits are immense works, all built of stone and concrete in their massive thickness, upon which neither time nor flood can make perceptible marks."[37]

At the beginning, selling the idea of electric power to the established textile manufacturers in the area was not always easy. Aside from up-front and short-run expense in changing equipment and procedures, many people remained frightened of electricity. An engineer for the Southern Power Company tried to sell one textile manufacturer on the idea of electric power. After listening for a quarter hour, however, the manufacturer ended the discussion by asserting, "You must be drunk or a damned fool if you think I will bring electricity into my mill to kill my people."[38]

To overcome such fear and skepticism and to encourage the mill owners to buy electric power, the Dukes invested heavily in a large number of textile mills in both Carolinas. Even before the launching of the Southern Power Company, the Dukes had taken the lead, and put up the lion's shares of the necessary capital, in establishing a large textile bleaching and finishing plant in Greenville, South Carolina. After 1905, however, their investments in the textile mills in the area around Charlotte, Spartanburg, and Greenville began to increase dramatically.

When James W. Cannon, the founder of what grew to be one of North Carolina's largest textile enterprises, approached J. B. Duke about investing in his mill, Duke sent word that if Cannon would "go to the site of one of our water power developments, say Great Falls for instance, and build a large plant, . . . he [Duke] would be glad to become interested, and largely interested with you." Duke thought the plant "should be a very large one, operating not under 100,000 spindles, and possibly 200,000 to 250,000 spindles." The mill might well produce "plain sheetings" and do so in such large quantities as to become known in the whole country as "the largest and best in this line of production." A vast amount of capital would be required for building on such a scale, but Duke thought "the money could be provided without much difficulty."[39]

Although Cannon did not go to Great Falls, another large textile mill did locate there, and the Dukes invested substantially in it. Later, James Cannon's son, Charles A. Cannon, asserted that Kannapolis, home of the extensive Cannon mills, could never have been built without the Southern Power Company. His mother had nervously inquired, "Now, are you sure that this electric power will be a success?" And his father had confidently replied, "Yes, Mr. Duke will make it a success."[40]

What was occurring in the textile factories of the Piedmont Carolinas in the early years of the twentieth century paralleled developments across the country. One historian asserts that electrification was probably "the most sweeping and complex technological change in American manufacturing over the past century. . . ." While the fundamental breakthrough occurred in the 1890s, the spread of electrification was held back, according to this historian, by a number of technical and economic factors, the most serious being "the general unavailability of cheap electricity." This problem was solved by the growth of the electric utility industry, a development that gathered full force after 1905. By 1914 utilities generated over half of all the electricity in the nation, and by 1967 over 90 percent. The historian points out that the emergence of the new, specialized industry with its central power stations was especially stimulating to the electrification of manufacturing plants. The special impact of electricity within manufacturing plants was to replace the coal-steam combination that had supplied most of the power requirements during the last half of the nineteenth century. The rise of electric power, then, helped hasten the demise of the steam engine.

The overriding reason for the adoption of electrical systems was the expectation—and then realization—of large cost savings. Electrification was the sort of technical advance that reduced all costs—labor, capital, and materials. The United States Commissioner of Patents had uncannily prophesied in 1849, when electricity was still primarily a scientific curiosity, that the belief was "a growing one that electricity in one or more of its manifestations is ordained to effect the mightiest of revolutions in human affairs." Yet a half century later one of those revolutions he had prophesied was underway in factories and workshops across the country.[41]

For a Tar Heel version of the transition described above, a visitor to the Great Falls plant in 1908 provides a colorful example. The visitor had worked as a general manager for a number of textile mills in Charlotte and vicinity, beginning in the 1880s. With extensive experience in the use of a variety of steam engines, he had finally succeeded in getting his "coal down to something like 2½ to 3 pounds per horse-power per hour." He also claimed to be the first man to start an electric-driven cotton mill in the Charlotte area, putting in "a turbine engine with generators direct connected, and this operated the mill until the Southern Power Company got [its] Great Falls Power ready for delivery." "But what about the steam engines?" he continued. "They have been with us for a long time but they must soon go. They have seen their day and served us well. . . . But I must now say farewell, old engine, I see your finish, I love you but. . . . you will be compelled to give place to the turbine water wheels and steam turbine with generators direct connected for the reason that they give a steadier, smoother, and more satisfactory speed." Hats off, he concluded, to Wylie, Lee, and their associates for "the wonderful work they have done at Great Falls. . . ." Now that "this great electrical development has passed the stage of experiment, you no longer have to back men into a corner and chain them there in order to talk to them about electric drives and lights." The day was not far distant, he prophesied, when men would be hunting for the representative of the power company to buy electric power "for all kinds of manufacturing purposes."[42]

Even before the plant at Great Falls was fully completed, the company began building another hydroelectric plant at Rocky Creek on the Catawba just below Great Falls and another at Ninety-Nine Islands on the nearby Broad River in South Carolina. The short-lived panic of 1907—what a later generation would call a recession—caused J. B. Duke temporarily to slow things down, but the Rocky Creek plant went into service in 1909 and Ninety-Nine Islands in 1910. In that latter year the company began construction of two small, strictly auxiliary coal-burning steam stations, one at Greenville, South Carolina, and another at Greensboro, North Carolina. More expensive to operate than the hydroelectric plants with their free "white coal," the steam plants were primarily for stand-by use and emergency situations.

As early as 1911, therefore, the Southern Power Company had linked together four hydroelectric plants—Catawba, Great Falls, Rocky Creek, and Ninety-Nine Islands—and two steam plants. In 1910 the editor of *Electrical World*, an important national trade journal, credited the Southern Power Company with "stimulating a whole population [in the Piedmont Carolinas] from a condition of former commercial and industrial apathy to an activity comparable... to that which characterized a new Western State."[43]

To credit the power company alone with awakening a "whole population" from a deep, southern slumber was something of an overstatement perhaps. A writer in *Electrical World* in 1914, however, was on more solid ground when he hailed "The Great Southern Transmission Network." He began by asserting that there had quietly grown up in the South "what is today by far the most extensive interconnected transmission system in the world." Observers in the electric utility industry had been aware for some years, the writer continued, that "splendid work" was being done in the development of the waterpowers of southern Appalachia and the Piedmont. But the linkage of the various networks into what already approximated, and one day would become, a united whole was "a comparatively new phase of the situation."[44]

This then-unrivalled "system of systems" referred not only to the Southern Power Company's network of generating plants with interconnected transmission lines, but also to the fact that J. B. Duke and W. S. Lee moved quite early to establish transmission links with other utility companies in the Southeast. Linkage with the fledgling Carolina Power and Light Company in eastern North Carolina and northeastern South Carolina came early (1912). The Georgia Railway and Power Company undertook the building of a large development at Tallulah Falls in northern Georgia. Anticipating a surplus of power, the Georgia company agreed readily to the Southern Power Company's offer to buy 2.5 million kilowatt-hours of electricity a month. In order to make the delivery to the Southern Power Company, a double-circuit, steel-tower transmission line had to be built in 1913 from Tallulah Falls to Easley, South Carolina, a distance of forty-five miles. The two companies split the cost.[45] This far-sighted arrangement would turn out to be critically important for the Southern Power Company when disaster in the form of unprecedented flooding on the Catawba struck in 1916.

Since transmission of electrical power was as important in the electric-utility industry as power generation, the Southern Power Company took pride in the fact that in 1909 it put into service the industry's first double-circuit 100,000-volt line. Stretching from Great Falls to Salisbury, North Carolina, and from there to Greensboro and Durham, it covered over 190 miles.[46]

Such technological breakthroughs were pushed by J. B. Duke and W. S. Lee, but others also played important roles in the company's early history. Charles I. Burkholder, a native of Illinois and engineering graduate of the University of Wisconsin, joined the Southern Power Company in 1906 at the age of twenty-four. Early professional experience with the General Electric Company gave him expertise in transmission line operation, and in Charlotte he was named as operating manager in charge of transmission lines, powerhouses, and substations. Like so many others in the company, he would choose to remain with it for his entire career, holding various higher positions until his death in 1948.[47]

The company's top leadership came not only out of engineering but also from the legal profession. A young lawyer who joined the company in its infancy, Norman A. Cocke, was a native Virginian who graduated from New York Law School in 1905 and soon, at age twenty-one in 1906, became an attorney for the Southern Power Company. He would one day head the company as well as play a prominent role in the civic and educational life of North Carolina.

Another young Virginian whose ability J. B. Duke spotted early on was Edward C. Marshall. A great-grandson of Chief Justice John Marshall, young E. C. Marshall began his business career at age eighteen (in 1895) with the Seaboard Railway in Portsmouth, Virginia. He soon took a job with the American Tobacco Company in New York. In 1907 J. B. Duke asked Marshall to become the auditor of the Southern Power Company, and he would subsequently hold several top-management positions before becoming president of Duke Power in 1949.

J. B. Duke enjoyed the company of hard-working, talented people like Lee, Burkholder, Cocke, and Marshall. The power company, however, was only one of his business interests in the first decade or so of the twentieth century. By 1905, when the Southern Power Company was formed, the giant American Tobacco Company, which he had headed since its organization in 1890, held sway over approximately four-fifths of the production in the United States in all lines of tobacco, except for cigars. With President Theordore Roosevelt on the warpath against "the trusts," the American Tobacco Company's turn came in 1907, when the federal government launched its anti-trust action against the Duke-led tobacco companies. (The United States Supreme Court would order the dissolution of the American Tobacco Company in 1911.)

Even before all that had begun, J. B. Duke led his company into a program of globalization that matched anything that the 1990s would come up with. Canada, Australia, Japan, China, and various other countries became prime markets for the South's famed bright leaf tobacco and the American Tobacco Company's products. After all, as early as 1889, young Duke had boldly announced that "the world is now our market for our product. . . ."[48]

In 1902, after a bold foray into England the year before to fight for a larger share of the tobacco market there, J. B. Duke decided that a wiser course of action would be to cut a deal: the American Tobacco Company and its affiliates would relinquish all of their business in Britain and Ireland, while the British-owned Imperial Tobacco Company agreed not to manufacture or sell tobacco in the United States, any of its dependencies, or Cuba. Then, to carry on the tobacco business in the world outside of the United States and Britain, the new British American Tobacco Company was incorporated under the laws of Great Britain with headquarters in London. The Duke-led companies received approximately two-thirds of its stock and the Imperial Tobacco Company the remaining third.[49] While J. B. Duke washed his hands of the United States' domestic tobacco business after 1911, he remained very much involved with the British-American Tobacco Company. The Southern Power Company, therefore, was not necessarily a top priority for J. B. Duke in the early days. Unforeseen events, however, would soon lead to its assuming a more prominent place in his life. And from the beginning it consumed an immense amount of capital.

Although the Duke brothers sometimes complained privately about the vast amounts of capital demanded by their power projects in the Carolinas, they never failed to provide the money as it was needed. On some occasions, they sold their stock holdings in other enterprises to plow more capital into the power company and its related enterprise. Norman Cocke later recalled one occasion when he was talking with J. B. Duke about some needed extensions. When Duke asked where the money was coming from, Cocke replied, "We're going to get it from you." "No you ain't," Duke stated. "I'm like the farmer who had a young steer that he wanted to break to a yoke, so he got a double yoke and he put his head through one side of it, and he put the young steer's head through the other. And the steer lit out and he ran all around the yard, and this fellow couldn't stop. He hollered to his son and said, 'Come here, Bill, and head us off, durn our fool skins!' That's what I want somebody to do for me," Duke laughingly concluded, "head me off."[50]

That J. B. Duke actually wanted to be "headed off" may well be doubted. He kept coming up with power-related projects in the Carolinas that soaked up capital like a sponge does water. To facilitate the industrialization that was his goal—and also to use some of the ever-increasing amounts of available electric power—Duke launched an electric interurban railway, the

Piedmont and Northern, in 1910–1911. The South Carolina branch of about 100 miles linked Anderson, Greenwood, Greenville, and Spartanburg with Charlotte; and the North Carolina division ultimately had about fifty miles of track, with the main line running between Charlotte and Gastonia, along with several shorter branch lines. Carrying both passengers and freight, the Piedmont and Northern, with its proud slogan of "A Mill to the Mile," became another important factor in the economic life of the region. The last of North Carolina's electric railways to switch to diesel engines (in the 1950s), the Piedmont and Northern had a proud, efficient history of its own until it was merged into the much larger Seaboard Coast Line railroad in 1969.

The Duke brothers, together with W. S. Lee and Edgar Thomason, a long-time official of the railroad, were the key figures in the building of the "P. & N." One example of the role of the Dukes came in 1913 while J. B. Duke was occupied in London as chairman of the board of the British-American Tobacco Company. B. N. Duke was left in New York to cope with the financial needs of the electric railway. When Lee informed B. N. Duke in June 1913, that work on the railway would require $250,000 in the next ten days, B. N. Duke advanced $100,000 and cabled his brother in London for $150,000. A few weeks later, when Lee needed additional funds and proved unable to arrange a loan in New York, B. N. Duke went to work on the original subscribers to the syndicate that he and J. B. Duke had organized to back the railroad and managed to get most of them to increase their subscriptions. "I feel very much gratified that we could handle the matter in this way," Ben Duke declared, "rather than borrow so much money at a time like this. I figure that we will get in say, $1,250,000 from these increased subscriptions. . . ." He concluded, somewhat wearily: "I hope no new and unexpected thing will turn up that will require additional money for I am about worn threadbare. I have had to pay out over a million dollars since January 1st for enterprises that I did not know the first of the year I would be called upon for." Over $600,000 of that, Ben Duke noted, had gone for power-company purposes, but he hoped that "the storm is now over and that we can take things more quietly."[51]

That particular cash crunch no doubt passed, but the capital needs of the power company were an on-going problem. The significance of the manner in which those needs were met should be emphasized, for the autonomy enjoyed by the Southern (and then the Duke) Power Company was rare among electric utilities. The expensiveness of so much of the machinery and equipment that were needed by the utility companies led most of them to pay for equipment with stocks and bonds of the company. In this manner, General Electric set up one of the early, major holding companies, Electric Bond & Share, in 1905. Even earlier, the Thompson-Houston company set up United Electric Securities to own the holdings of companies that had paid for equipment with securities. There were other reasons that led to the emergence of the electric utility holding companies, but the main point is that by 1932, the eight largest holding companies controlled 73 percent of the investor-owned utilities.[52] Not only were crucial decisions concerning the utility companies made in northern cities quite distant from the places where the utilities actually operated, but also the holding companies were not regulated. That is, they were not regulated until important New Deal legislation was enacted in 1935. As for profits, the biographer of Samuel Insull, a major figure in the history of the electric utility industry, has succinctly noted that "... financing utilities was many times as profitable as running them."[53]

Commonwealth and Southern, which became one of the largest utility holding companies in the nation, owned at one time the common stock of five northern utilities and six southern companies—Alabama Power, Georgia Power, Gulf Power of Pensacola, Mississippi Power, South Carolina Power, and Tennessee Electric Power. All of the southern group (except the Tennessee company) had previously been owned by another holding company, Southeastern Power and Light. An official of South Carolina Power relates that when he and one or two associates took their company's annual budget to the president of Commonwealth and Southern in New York for review, the president—apparently not too concerned about intra-company relations—would let them cool their heels in an outer office for two or three days before granting an audience.[54]

The Virginia Electric and Power Company (VEPCO) was a subsidiary of a large holding company from its start until 1947, when it became an independent, investor-owned utility. Likewise, Carolina Power and Light Company escaped from a New York holding company and had its stock listed on the New York Stock Exchange in 1946.[55]

The Southern (Duke) Power Company's unusual autonomy was, of course, merely one source of the company's efficiency, reasonable rates, and high morale. Enough has been said, however, to show why the rumor that began to spread around 1912 that J. B. Duke might transplant himself permanently to London inspired deep consternation in the minds of business-minded Carolinians in the Piedmont.

That J. B. Duke lost no love for either the trust-busting policies of Theodore Roosevelt or much of the reform agenda of President Woodrow Wilson's New Freedom was a fact. (Duke's all-time favorite president was none other than William McKinley, Republican apostle of "protectionism and prosperity.") It was also true that, after repeated visits to London beginning in 1901, Duke grew to appreciate the fact that life in Edwardian England, with its greater formality and marked privileges according to wealth and class, held certain attractions. Moreover, he believed, correctly, that any Englishman who had built up a great industry and helped develop a huge foreign market for a major agricultural product such as tobacco, would have been, at the least, knighted—and not vilified as a "robber baron" and "trust ogre."

Accordingly, in 1914 J. B. Duke leased one of the great mansions of Mayfair in London's fashionable West End, with an option to purchase. The outbreak of World War I later in that year, however, ended any notions that Duke may have had about owning even a part-time residence in England and sent him scurrying back to America.[56] Before all that transpired, however, a prominent South Carolina businessman, worried about the possible allure of London for J. B. Duke, wrote him a letter that inspired an unusually articulate and significant response. (Not given so much to words, J. B. Duke acted on the principle that what he did was much more important than what he might say.) The businessman, having missed seeing Duke in either Charlotte or New York, explained that he wanted to express his "very sincere hope that your residence abroad is going to be only temporary." He continued: "We need you in the United States, and particularly in the South." The great obstacles that had to be overcome in the development of the South's "wonderful latent resources" simply required "men of the broad constructive abilities which you possess to so marked a degree." Duke's personal interest in "our electric power and transportation matters" and in the "direct introduction of our Southern goods to the Orient" had been "highly gratifying." The South Carolinian concluded by declaring that he wished to assure Duke that, "I feel very keenly the loss that it will mean to us if your personal interest and activities should be devoted elsewhere."

J. B. Duke, who generally delegated the task of responding to letters to his executive secretary, could hardly ignore such a gracious tribute as the South Carolinian had paid him. Accordingly, he personally replied in a rare articulation of his priorities and inner thoughts. "It is true," Duke noted after expressing thanks for the letter, "that it has become necessary for the protection of my own interests that in the future I devote a considerable portion of my time in the British Empire and the Continent of Europe." That fact alone, however, "does not mean that I have the in the slightest degree lost interest in my native land and its people, especially that part lying south of the Potomac and embraced in the States of North and South Carolina." Duke continued, "I have always felt that this particular region had, by reason of its natural resources and climatic conditions, the possibility of becoming one of the most favored and desired spots on this Continent or elsewhere." Duke believed that the "one thing necessary... to bring this about is capital directed by intelligent effort." He went on to say:

> With this end in view I have within the past five years invested and caused to be invested in the two states above named, approximately twenty-five million dollars. I have invested this money along lines calculated to do most for the general good of the two states and the prosperity of the people; that is, in the development of water powers running to waste and in the improvement of transportation facilities. Those investments can be made profitable and even exceedingly attractive if we can secure the co-operation of your people (in supporting them with their business); but without this, they are doomed to failure....

J. B. Duke concluded his rare letter with a combination of sentiment and mild threat. He could never, he declared, "lose a sentimental interest in the land that gave me birth; but my financial interest in the future must necessarily be limited by the appreciation shown and the co-operation afforded those investments already made.... To sum up briefly, my future investments in your state and elsewhere in the South will depend upon the encouragement I receive at the hands of your people and not upon my place of residence or the investments I have made in foreign countries."[57]

People in the Piedmont need not have worried. J. B. Duke was one Tar Heel who had never totally left home, and events would soon transpire to bring him closer to his native region than ever. His beloved power company, however, was headed for a series of unanticipated and difficult challenges—an unprecedented flood on the Catawba, then a dangerous political battle, and finally a cruel, summer-long drought that, in a way, was as harmful to the power company as any flood.

CHAPTER II

EXPANDING A TRAIL-BLAZING SYSTEM, 1911–1925

While the men who launched the Southern Power Company did have a great organizing vision of a series of inter-connected hydroelectric plants on the Catawba River, there was no master blueprint for what would eventually become the Duke Power Company. That is to say, much that happened was not foreseen or planned but just occurred willy-nilly in response to unanticipated circumstances.

As mentioned earlier, J. B. Duke's primary purpose in bankrolling the power company was to spur industrialization in the Piedmont Carolinas through the growth of textile manufacturing. The Southern Power Company originally had no plans for supplying electricity to residential customers. Yet early on, many of the textile manufacturers insisted that electric power should be supplied not only to the mills but also to the residents of the mill villages. Thanks to the long-distance transmission of electricity, textile mills no longer had to be clustered around power sites or concentrated in cities but could be scattered throughout the countryside as the owners might choose. This was one reason why industrialization in the Piedmont Carolinas did not immediately result in the urbanization that had occurred earlier in New England, for example, and even earlier in Britain.

The Southern Power Company, at any rate, quickly found itself serving a larger and more diversified array of customers than had been foreseen. Moreover, the privately-owned electric power and light (and gas and/or water) companies that had been organized in so many cities and towns across the Piedmont (as well as the rest of the country) in the 1880s and 1890s quickly realized that the Southern Power Company could produce electricity that was both cheaper and more reliable than that provided by the local utility. Consequently, the Southern Power Company began first supplying bulk power to the municipal companies and then gradually buying many of them out altogether. This was a process that accelerated in the 1920s and during the depression of the 1930s but one that continued well into the latter half of the century.

Along with the growing number of residential customers, another unforeseen aspect of the business turned out to be the sale and popularity of electrical appliances. Just as electricity transformed the American factory so would it gradually alter drastically private homes and the standard of living. One of the first appliances to be marketed was the electric iron. Its obvious advantages over the stove- or fireplace-heated irons, especially in hot weather, made it relatively easy to sell, so one early salesman for the power company strapped some electric irons to the back of his bicycle and took them to customers for a two-week trial.[1] Toasters, percolators, and vacuum cleaners came along soon after the electric iron.

An historian who has focused on the social impact of electrification points out that once homes had hot water heaters, bathing became more frequent, doing laundry at home required less work, and washing dishes was easier. Accordingly, standards of cleanliness and the frequency of many tasks tended to increase. The electric light, which was really the first appliance, came to be regarded as a necessity in most of the nation long before central heating, indoor toilets, or the telephone.[2] Electricity, in short, was an enabling technology that was not always noted. "It quietly became central to the functioning of the modern city," the social historian observes, "to the 'industrialization' of the home and the modernization of the factory, and [finally] to the improvement of the farm."[3]

At first, Southern Power and other early utility companies may not have paid much attention to the fact, but the ever-growing number of residential customers and their ever-increasing array of electrical appliances would eventually become an important consideration in "load management." That is, it became crucially important to the Duke Power Company in the early 1930s when the depression sharply curtailed electricity usage by textile mills, that sales to residential customers continued to rise. Even before that happened, it was economically advantageous to the power company that, while streetcar systems and factories used more electricity during daylight hours, many residential customers burned their lights and used certain appliances at night. (Later on air-conditioning would complicate the pattern.) Long before worries about peak loads—the times of maximum electricity usage—became so critical in the 1970s, the matters of load management and balancing the load were important considerations for those utilities like Duke Power that strove for the most efficient and economical use of their generating plants.

Although home electrification jumped from 10 per cent of the total number of homes in 1910 to 70 per cent in 1930, the early electrical appliances were not always a source of unmitigated pleasure.[4] A young salesman with the Alabama Power Company relates that, after knocking on countless doors in 1928, he finally managed to sell a "Whirl-Dry" washing machine to a "buxom housewife"—and he welcomed his $6 commission. Later in the day, however, one of the lady's neighbors called to report that the washing machine had his customer pinned to her kitchen wall. It seems that when the drying cycle was reached, the cylindrical copper tub, revolving at high speed, gave the machine a tendency to "walk" across the floor. "I retrieved the machine," the salesman explained, and "the company retrieved my six dollars. I sold electric ranges and refrigerators with somewhat better success."[5]

Finding the Southern Power Company involved in operating streetcar systems in Charlotte as well as various other cities throughout the Piedmont and in the ever-growing retail distribution of electricity, J. B. Duke and his associates decided in 1913 that a separate organization focused on retail operations was needed. Accordingly, they created the Southern Public Utilities Company and named as its first president Zebulon Vance Taylor. He had resigned from the presidency of the Greensboro (North Carolina) Electric Company in 1910 to join the Southern Power Company. Upon Taylor's death in 1921, E. C. Marshall became president of the Southern Public Utilities Company and remained in that position until the company was merged with the Duke Power Company in 1935.

Even before the organization of the Southern Public Utilities Company, the power company acquired its first subsidiary in 1910, the year in which J. B. Duke succeeded W. Gill Wylie as president of the company. Duke and his associates realized that the conversion of textile factories from steam to electricity required the purchase of extensive and expensive new equipment. Moreover, the power company itself could use a purchasing agent. Consequently to facilitate the above-mentioned conversion and better serve the company's own needs, the Mill Power Supply Company was organized. It would become a substantial enterprise, one that tried in a variety of ways, such as extending low-interest loans and generous terms of credit, to encourage mill owners to use electric power.

Both the generation and transmission components of the company grew steadily. In addition to the two auxiliary steam stations that had already been built (at Greenville, South Carolina, and Greensboro, North Carolina), two more were soon added. The one at Mt. Holly, North Carolina, came on line in 1913 and the Eno River steam plant near Durham began operating in 1915. Durham, it might be noted, then marked the eastern limit of the area served by the Southern Power system, though at a much later date both the Research Triangle Park and the Chapel Hill area would be added.

In 1913 the company began constructing another hydroelectric plant on the Catawba, this time at Lookout Shoals about eleven miles west of Statesville, North Carolina. With a generating capacity of 33,000 horsepower (18,720 kilowatts), the Lookout Shoals plant began operating late in 1915 and produced about three times as much electricity as any one of the small steam plants in the system.

Just as everything seemed to be going so well for the company, disaster struck in July 1916. Unprecedented rains falling on the eastern slopes of the Blue Ridge Mountains in western North Carolina caused the trouble. The rains came in two stages: first, in early July the remnants of what was then called a tropical cyclone passed inland over the Gulf coast into northern Alabama, eastern Tennessee, and the western Carolinas. These rains thoroughly soaked the earth and filled the streams and rivers. Then on July 15–16, the remnants of another tropical cyclone moved inland from the Atlantic coast and in a twenty-four hour period dumped over nineteen inches of rain on the mountainous area where the Catawba River originates.[6]

"Greatest flood since Noah," declared old-timers residing near the Catawba. A young employee who had just been hired in the Dukes' New York office, which maintained constant and direct links of communication with the power company and all its plants, reported that a stenographer emerged from an inner office with a dazed look. Then he announced woefully that they had all best look for new jobs because the "rains have been so heavy down there... at the Catawba river, that the company's washed the hell down the river".[7]

Governor Locke Craig reported to the North Carolina legislature that, "There has been no such swell of waters since the country was settled by white people.... The sides of mountains were torn loose. The valleys were flooded by raging torrents. Trees, crops, buildings, roads, railroads and the land itself were swept away.... The rich bottom lands of the Yadkin, the Catawba and tributaries were turned to desert wastes."[8]

On the Southern Power system, the Rocky Creek hydro station just below Great Falls suffered the greatest damage. When river debris settled on the plant's roof, it caved in and the raging river swept through the plant. The original Catawba plant at India Hook went out of service on July 16, 1916, and could not resume service until December. A number of other plants were knocked out for periods ranging from six days to two months.[9]

The power company's loss, both in equipment and from lost kilowatt hours of service to customers, ran into millions of dollars. But by purchasing power from every available source—and the earlier linkage with the Georgia Railway and Power Company now proved invaluable—and by carrying dangerous overloads on its own plants, the company slowly struggled to its feet. "Every useful and available member of the organization was put to work," a veteran employee recalled, "and sleep was at a premium until the restoration . . . of the plants was completed."[10]

Although J. B. Duke well knew he could not check the rains from the sky, he rushed south, alarmed as well as angry about the large losses. Traveling around from plant to plant inspecting the damage, he no doubt made frequent use of his favorite expletive, "I'll be dinged." E. C. Marshall, in whose home Duke frequently stayed while in Charlotte, finally confessed: "Mr. Duke, I don't blame you for being mad. If that were my money floating down the river, I would be mad too. . . . We just don't know how to cope with the river and the flood." Duke grunted his assent to the explanation but added that he meant to find out how to cope with such problems and that he expected the others in the company to apply themselves promptly to the same task.[11]

Finding out how to cope with such a problem was precisely what W. S. Lee, C. I. Burkholder, and other first-rate engineers with the company quickly proceeded to do. The solution they came up with proved to be perhaps the most impressive and imaginative single project achieved on the Catawba up to that point: they would build a series of three massive interlinked dams on the headwaters of the Catawba and create a vast reservoir behind the dams. Not only would the dams help to avoid future flooding such as had occurred in 1916, but the reservoir could also be utilized to increase stream flow on the lower Catawba during the low water of dry spells. The project would prove to be a giant step in the taming and enhancement of the Catawba and thus of the company's harnessing of the entire river.

Construction on the Bridgewater plant and reservoir, which would be named Lake James (in tribute to James B. Duke), began in August 1916, so obviously no time was lost after Duke's trip south and pronouncement that he wished prompt action. W. S. Lee and Richard Pfaehler, a younger engineer who worked with Lee on the project, were understandably proud of what had been done at Bridgewater and wrote several detailed accounts of the project for the leading engineering journals.

The Bridgewater powerhouse was located in mountainous Burke County at the small town of Bridgewater, about eight and a half miles west of Morganton, North Carolina. Because the tracks of the Southern Railway ran alongside the Catawba and east of the town of Bridgewater, the engineers realized that to build a high dam across the Catawba alone at that point would be impracticable. Therefore they hit on the solution of building three separate dams, all within a radius of two and a half miles from the railroad station at Bridgewater—one on the Catawba itself and two on tributaries that fed into it, Paddy Creek and the Linville River. Since topographical and geological conditions at the three sites precluded the use of masonry dams, they built earth dams, one of which was said to be among the highest in the world at that time.

The drainage area of the Catawba tributary to the Bridgewater storage reservoir was 370 square miles and the average rainfall was from 55 to 60 inches. The headwaters of the Catawba come from the steep slopes of the Blue Ridge and Linville Mountains, and heavy rains cause frequent flooding. Using records compiled by the Southern Power Company, the engineers estimated that on July 16, 1916, the maximum discharge of the Catawba into the area that would become Lake James corresponded to 150 cubic feet per second per square mile of the drainage area, which was equivalent to a depth of water of 5.58 inches over the whole watershed. Therefore, the engineers designed a spillway capacity corresponding to twice the record discharge of July 16, 1916, over the entire watershed, before the reservoir level would reach the top of any of the impounding earth dams.

When full, the water surface of the entire reservoir (which had three arms formed by the valley of the Catawba and Linville rivers and Paddy Creek) was at elevation 1,200 feet covering an area of 6,509 acres. As long as the water surface was above elevation 1,167, all the water stored in the reservoir would be discharged through the turbine at Bridgewater Station and produce 20,000 kilowatts of electricity. The plan called for stopping the operation of the power station as soon as the water surface of the Paddy Creek and Linville arms of the reservoir fell below elevation 1,150. Stored water between elevation 1,150 and elevation 1,105 would be utilized, however, to make up a portion of the deficiency in the flow of the Catawba below Bridgewater by opening relief valves. Altogether, the project represented a remarkable and highly creative feat of engineering.[12]

The Bridgewater plant went into operation in 1919, but the critical test for the flood-control aspect of the project would come in 1940. In August of that year came the highest flood on record for the upper reaches of the Catawba, a flood even worse than the one in 1916. As a result of the operation of the Bridgewater reservoir, however, the flood stage at Morganton in 1940 was 14.5 feet below the 1916 mark. At other points on the river, the 1940 levels were similarly reduced.[13] While the plan for releasing water from Lake James to increase the flow of the lower Catawba during dry spells proved to be highly effective and beneficial to the company, not even that arrangement would prevent a desperate situation for Duke Power when rains refused to come in the long, hot summer of 1925.

At the same time the Bridgewater-Lake James project was being built by the Western Carolina Power Company, a subsidiary of the Southern Power Company, J. B. Duke and his associates organized another company, the Wateree Power Company, to build, the Fishing Creek and Wateree plants, the latter to become the largest plant on the system up to that point. One reason for the organization of the Wateree Power Company was that the Southern Power Company did not have the power of eminent domain. While the company owned virtually all of the property necessary for the two developments on the lower Catawba, there were a few small properties in the area which it had not acquired. The company later explained to the Commissioner of Internal Revenue in Washington: "A failure to acquire these properties, or a defective title which had not been discovered, or an error in determining the water level might, in the case of a company which did not have the power of Eminent Domain, result in an injunction proceeding brought by some small property owner, and to avoid this the Southern Power Company caused the Wateree Power Company to be incorporated by a special Act of the General Assembly of South Carolina in 1909. This has customarily been done in South Carolina by various other power companies."[14]

One should note that J. B. Duke, W. S. Lee, and their associate believed in carefully looking ahead and acquiring land and water rights as early as possible, long before the public announcement of a project. The Wateree Power Company was incorporated in 1909, but construction on the Wateree hydro-station did not begin until November 1916. In May 1917, moreover, J. B. Duke arranged to have the Wateree *Electric* Company incorporated in New Jersey. This was important in the history of the company, for the Wateree Electric Company gradually became the owner of a number of power companies that J. B. and B. N. Duke had organized (Great Falls Power Company, Wateree Power Company, etc.). In November 1924, it was the Wateree Electric Company that changed its name to the Duke Power Company. And while the Southern Power Company (begun in 1905) kept its name and separate identity for a few more years, in December 1927, it too was merged into the Duke Power Company.

Fortunately for W. S. Lee and the growing cadre of first-rate engineers recruited by the company, they did not have to bother themselves about corporate structure and names. The engineers, and the hundreds of those who worked with them, got generating plants, dams, and reservoirs built with remarkable alacrity. Work on the Fishing Creek hydro station, located on the Catawba about two miles north of Great Falls, South Carolina, began in April 1915, and the plant, capable of producing 36,720 kilowatts, went into operation in November 1916.[15]

No sooner was that substantially completed (and while work on the Bridgewater project was also underway) than work on the largest plant so far built on the Catawba began in February 1917, and operations at the plant began in September 1919. The Wateree hydro station, about eight miles northwest of Camden, South Carolina, was the farthest downstream Duke plant on the Catawba-Wateree River. (The name of the river changes to Wateree about sixteen miles upstream where Wateree Creek feeds into the Catawba.) Big for its time, the Wateree plant cost $5¼ million and its installed capacity of 56,000 kilowatts was more than one-third of the Duke system's total capacity at that time.[16]

Several engineers who would become major figures in the history of Duke Power and spend the remainder of their careers with the company signed on during this period of major construction. (It went on, of course, even as President Woodrow Wilson in April, 1917, led the nation into World War I.) One of these engineers who would achieve a national reputation in his field was David Nabow. Born in Russia to Jewish parents in 1895, young Nabow emigrated with his family to the United States as a child. Graduating from New York's famed Stuyvesant Technical High School in 1912, he won a scholarship to attend Columbia University's engineering school, from which he graduated in 1916. Just how this brilliant young engineer came to the attention of the Southern Power Company is not known, but he began his first full-time employment with the company in July 1916. Working at first on problems connected with hydroelectric developments—among other things, he perfected and patented "the Butterfly type of headgates" used in all the subsequent Duke Power hydro developments—he would later achieve a national reputation for his work in the design of steam-electric plants.[17]

Another young engineer who joined the company in January 1918, was Charles T. Wanzer, who was born in Ithaca, New York, in 1889. After graduating from Cornell University as a civil engineer in 1913, Wanzer held a number of positions before he headed south and helped first to build the Wateree plant. He was destined to rise through the ranks and, like David Nabow, eventually became a vice president of Duke Power and chief engineer before his retirement in 1962.[18]

Not only did J. B. Duke enjoy the company of these able engineers and the others associated with the company in the Charlotte headquarters, but he also particularly liked to be around large construction projects involving water, stone, and other materials. Consequently, he made frequent trips to Charlotte and from there to construction sites at Bridgewater, Wateree, and elsewhere. In Charlotte he had no office of his own but, as Norman Cocke later recalled, would arrive at the power company's offices around 7:30 or 8:00 a. m. Interested in the loads at different plants, he would often remain for long periods in the central operating office. Sometimes he would just drop in on W. S. Lee or Cocke or Marshall. (If the recipient of these unscheduled visits was busy, one has to wonder if prolonged conversation was always genuinely welcomed.)

Some days J. B. Duke would have a simple picnic lunch prepared so that his valet-chauffeur, Fred Crocker, whom he had acquired in England, could drive him and one or two others out to one of the power plants or construction sites. One young engineer, newly employed by the company, met Duke with trepidation, for the word had been passed around that "Mr. Duke was a dangerous man to talk to, that he remembered everything you told him, and he might bring it out two or three years [later] in an embarrassing way." Some time after the introduction, Duke invited the young man to ride out with him for an inspection of a new coal-burning steam plant. After the nervous engineer managed to have someone telephone a heads-up alert to the plant, he and J. B. Duke set out in the latter's old chauffeur-driven Rolls-Royce. As the car topped a hill near the plant and the engineer saw that both stacks were clear with no black smoke coming out, he felt one spasm of relief. But as Duke's actual inspection of the plant began, the engineer's tension mounted. After going through turbine, pump, and boiler rooms, questioning all along, Duke walked outside and, to the engineer's great consternation, proceeded to walk into an opening at the base of a chimney, where soot was gathered with a small electric machine. The area turned out to be perfectly clean. Returning to the car and sharing sandwiches and a thermos of coffee with the engineer, Duke looked straight at him and said, "You have a fine plant here. I like the way you run it." As the young engineer began to glow, Duke added, "Why don't you run the others like this?"[19]

While the dam and power plant at Wateree were being built between 1917 and 1919, J. B. Duke went there frequently, often visiting in the home of the resident engineer, A. Carl Lee, younger brother of W. S. Lee. Smoking the small, mild cigars that he liked, Duke would sit in a large rocking chair after dinner, occasionally held the Lee's baby daughter, and talk about his own young daughter, Doris, who had been born in November 1912. Duke told Mrs. Lee that on one occasion he had been emphasizing to Doris that he had always been a Methodist, her grandfather had been a Methodist, and that he hoped she would be one. When Doris looked up at him and asked, "Daddy, what's the difference between a Methodist and a Presbyterian?" Duke had to do some quick thinking. "Doris, it's time for you to go to bed," he announced. "You go to bed and tomorrow we'll settle all that." The next day Duke hastily sought out a Presbyterian minister in Charlotte, got informed on a few of the fine points of theology and church history, and then tried to explain as best he could to his inquisitive young daughter.[20]

Duke worried about a public health problem that developed during and soon after the construction at Wateree and that led to an important addition to the power company's staff. Plagued by an out break of malaria in the construction worker's camp and then in nearby York County, South Carolina, J. B. Duke reportedly telephoned the Surgeon General of the United States for help and was advised to seek the services of Dr. Frank Moon Boldridge. A native South Carolinian, Boldridge had received his medical degree at the Medical College of Virginia in Richmond after graduating from Wingate College. A specialist in public health, he served with various divisions of the United States Army during World War I and after the war worked in the School of Tropical Medicine in Liverpool, England, where he specialized in malaria control.

Subsequently, Boldridge joined a group of United States Public Health officers on a mission to fight malaria in various Latin American countries, including the Panama Canal Zone. After that duty, he went to work for the South Carolina State Board of Health.[21]

In response to J. B. Duke's request, Boldridge began working as a consultant to the power company in 1921. Impressed by Boldridge's quick mind and determination, J. B. Duke in 1923 persuaded him to accept a full-time position with the company and take charge of mosquito control, sanitation, and municipal water supply. Boldridge would remain with Duke Power until his retirement in 1962 and would play a large role in the company's ever-expanding health and environmental program. From a one-person staff in the early 1920s, the company's environmental program would grow to include over 200 scientists and specialists headquartered in state-of-the-art laboratories on Lake Norman by the closing decades of the twentieth century.

As sound as J. B. Duke's judgment usually was, he was by no means infallible. One venture in which he invested considerable capital and energy failed rather dismally, and that was his attempt to link his power plants on the Catawba to the vitally important and lucrative manufacture of fertilizer. Since the story is long, highly complicated, and only tangential to the history of Duke Power, no attempt will be made to retell it fully here, and, at any rate, it has been recently and well developed elsewhere.[22]

Duke's interest in the matter arose from the fact that at certain times when there were good rains and high water (but not too high) on the Catawba, his generating plants were from an early date capable of producing more electricity than there was a market for. Electricity, unlike textile or many other commodities, cannot be stored, and waste bothered J. B. Duke. He liked to remind his engineers and associates in the power company that before the plants were built, it was just water that flowed down the Catawba-Wateree to the ocean. But once the plants were installed, water wastage represented a loss in dollars.[23]

By 1908, J. B. Duke, working closely with W. S. Lee, launched an intensive and far-reaching search for an electrochemical or electrometallurgical (i.e., aluminum) use for his surplus electricity. After considering various other possibilities, Duke and Lee settled on the fixation of atmospheric nitrogen for use in fertilizers and explosives as the preferred project for the Southern Power Company.[24]

After considerable investigation, J. B. Duke purchased the United States or North American rights to several European patents and built an experimental plant on the Catawba south of Charlotte and just above Great Falls. (The plant site was named Nitrolee because work related to "nitrogen" was to be done there and W. S. Lee supervised the construction of the plant.)

Beginning operations in May 1911, the Nitrolee plant encountered problems. As the Austrian chemical engineer whom Duke had employed to help design and run the plant later testified, perhaps the principal problem was that surplus power was only sporadically available for the plant's operation. That meant that no regular schedule of production could be set or firm contracts for the sale of nitric acid negotiated. Because of these complications and other developments, the Nitrolee plant was shut down in 1916.[25]

Not to be deterred, J. B. Duke set out to find other water power sites where sufficient and cheap electricity could be produced and then used in the manufacture of nitric acid for fertilizer or gunpowder. Thus it was that in September 1912, J. B. Duke—accompanied by B. N. Duke, W. S. Lee, George G. Allen, a younger business associate who was destined to play a large role in all of the Duke-related enterprise, and various others—set out on a train trip to the Pacific Northwest in search of likely water power sites.

After the trip had been planned, J. B. Duke let Thomas L. ("Carbide") Willson, a pioneer in the electrochemical field, persuade him to change the itinerary and visit Canada's Saguenay River, a tributary of the St. Lawrence in a then-undeveloped, thinly-populated, and rugged region about a hundred miles north of Quebec City. Leaving their private railway cars upon arrival at the town of Chicoutimi on the Saguenay, J. B. Duke, Lee, and others eventually took to a "duck boat" to go up the river as far as they could; then they had to walk up trails along the bank of the river for two miles or so. Fed by Lake St. John, which covered an area of some four hundred square miles, the Saguenay dropped over three hundred feet in the approximately thirty-seven miles from the lake to Chicoutimi. J. B. Duke and others clambered over the rough terrain and gazed on portions of the mighty river as it broke and surged through the gorges. Duke and Lee immediately saw that the enormous potential of the entire run of the river, with Lake St. John as the reservoir, could be comprehensively developed, and Duke declared, "Lee, I'm going to buy this."[26]

Buy it Duke did, though the process of acquiring the two major power sites on the Saguenay—the upper one near the outlet of Lake St. John at Isle Maligne and the lower one twenty-one miles down the river at Chute-à-Caron—required over two years. Then for more than three years Duke's chief legal counsel after 1913, William R. Perkins, and other lawyers and agents struggled to purchase for the Duke brothers the vast areas of land required for the water rights of the enormous projects that were envisioned.

Just as J. B. Duke had to help build the market for his electricity produced on the Catawba, so even more so was there the necessity of finding purchasers for the electric power that Duke and Lee hoped to produce on the Saguenay. Promising negotiations between Duke and the E. I. DuPont de Nemours Powder Company of Delaware began in April 1913. The DuPonts, in need of steadily increasing amounts of nitric acid, were glad to explore the possibilities of joining forces with J. B. Duke to build generating plants and factories on the Saguenay.

Contemporaneous with all these developments, J. B. Duke made another move looking toward entering the fertilizer business, and doing so ultimately on the Saguenay. In addition to acquiring a power site on the Saguenay from "Carbide" Willson, Duke gained title to Willson's patents for fertilizer manufacture. In 1914, therefore, Duke authorized the construction of a plant in Mount Holly, North Carolina, to demonstrate on a commercial scale whether phosphoric acid could be competitively produced by electric furnaces. That plant was up and running by November, 1914, but, while all these ventures would ultimately lead to J. B. Duke's having a controlling interest in the American Cynamid Company, they were destined not to have any permanent impact on the development of the Saguenay or on the Southern Power Company.

Duke's promising negotiations with the DuPonts and others hit a roadblock when he failed to secure permission from the government of the Province of Quebec to raise the level of Lake St. John. For J. B. Duke and W. S. Lee, a vital part of the plan for harnessing the Saguenay had been, from the beginning, to impound the lake, thereby increasing the available "head" or fall at Isle Maligne and allowing the water stored in Lake St. John to be used to minimize seasonal fluctuations in the flow of the river. By early 1916, Duke had proven unable to win the necessary governmental clearance to raise the lake level, the DuPonts pulled out, and the whole Saguenay project had to be put on hold until the early 1920s.

Despite the enormous scope and economic significance of the Saguenay project, not only for the Dukes but also for an undeveloped and capital-hungry region of Canada, J. B. Duke's long-range interest in hydroelectricity remained focused on the Piedmont Carolinas. From about the time that W. R. Perkins became Duke's legal counsel in late 1913—and probably sooner—J. B. Duke had begun to turn over in his mind a plan for the large-scale philanthropy that he hoped to implement some day. The plan would involve J. B. Duke's giving a substantial portion of his stock in the power company to a charitable trust and then closely linking that trust to the actual management and operation of the power company. Considerable water would have to go under the bridge—or, more accurately, through the turbines—before J. B. Duke would have his plan fully worked out and ready to be announced. That he began contemplating a Grand Design for perpetual philanthropy in the Carolinas at least by 1913, however, is a well-documented fact.[27]

Giving concrete expression to J. B. Duke's intensifying interest in the power company and his native region was his decision in 1919 to acquire a home in Charlotte. His fondness for the business-minded and ambitious city had grown steadily since the establishment of the Southern Power Company in 1905, so he purchased a house in the new, close-in suburban section known as Myers Park and near homes of the E. C. Marshalls and other officials of the power company. Remodelling the house extensively and adding to and beautifying the handsome grounds around it, Duke installed a fountain on the grounds of his home that threw a jet of water more than eighty feet in the air. The house itself was spacious and comfortable rather than pretentious, but the fountain and grounds were Duke's special joys.

Always a lover of trees, he admired the stately white oaks on his property and especially liked to bring his family to Charlotte in the early spring when, as he put it, the white oaks "bloomed." (They did not actually bloom but budded out in a showy manner.)[28]

For all of Duke's keen interest in the power company, there was one bothersome thing: it was making too little money. Duke's closest associates in the enterprise were struck by the fact that he took neither salary nor expenses from the company during the many years he was so closely involved in it. More significantly, Duke was not interested in his and his family's receiving dividends from the large blocks of stock they held in the various companies that made up the Southern Power system. Every cent that the company cleared got plowed right back into additional power plants, transmission lines, or acquisitions of the Southern Public Utilities Company. In 1911, for example, the Duke's executive secretary informed B. N. Duke that the company had earned enough for dividends "but that it seems to be the same old story, namely that we have been spending so much money for extensions and new acquisitions that we have not sufficient money in the bank to pay the dividends without cramping ourselves." The executive secretary had taken "the matter up with Mr. J. B. Duke and his decision was that we should not attempt to pay the dividends on the 1st of April."[29]

Rather than enjoying dividends, the Dukes scrambled to find the necessary capital for the enterprise. In the spring of 1917, Ben Duke, his wife, and his two children had loans out to the Southern Power Company and a related company that totalled $1,020,000. By August 1917, Angier Duke, Ben Dukes's son, was selling the family's American Tobacco preferred stock as well as other stocks in order to gain additional funds for the power company.[30]

The 1916 flood dealt the company a costly blow, obviously, and that combined with the inflation that hit the nation's economy during and immediately after its participation in World War I put the company in a tight situation. It had made long-term, low-rate contracts with the owners of textile mills and other large consumers of electricity. When the company moved in 1919 to attempt to strengthen its financial position, that would have long-range consequences, both for the company and its future relationship with the North Carolina Corporation Commission,the body to which the legislature had made an ambiguous grant of power to regulate public utilities in 1913.

The problem of restraining the excesses of free-enterprise capitalism loomed large in the early twentieth-century United States. There were quite a few democratic socialists in the nation before World War I, people who believed that government ownership of the means of economic production could and should be established within a framework of political democracy. The great majority of Americans, however, remained wedded to the historic system of private property and free enterprise capitalism. When the electric utility industry began to develop in the late nineteenth and early twentieth centuries, therefore, it seemed perfectly proper to most Americans that the gigantic and costly task of electrification should be left to private capitalists and entrepreneurs like J. B. Duke, J. P. Morgan, and others.

Yet the Southern Power Company, like many other electric utility companies, was launched as the Progressive Era's reformers, led by President Theodore Roosevelt, began to have great influence and power. The unfettered capitalism favored by the likes of Roosevelt's predecessor, President William McKinley—J. B. Duke's all-time favorite president—and other Old Guard Republicans now confronted new challenges posed by reformers on the federal and state level. They called not for government ownership but for regulation to protect the public from the alleged excesses and abuses of unrestrained capitalism. The idea of municipally-owned electric utilities also gained considerable momentum around the turn of the century.[31]

As for the electric utility industry, Samuel Insull was one of the first to propose that it could function best as a natural monopoly in a specified area and should be regulated by a state agency that would fix rates and set service standards. A former associate of Thomas A. Edison who became the head of the Chicago Edison Company, Insull advanced his idea in 1898, but it did not gain immediate acceptance. By 1907, however, both the National Electric Light Association and the National Civic Federation, an influential organization composed of national business leaders and others, had endorsed Insull's proposal. In that year, three states established regulatory agencies, and by 1916 thirty-three states had them.[32]

North Carolina got on the Progressives' regulatory bandwagon in 1913, when the legislature, in response to Democratic Governor Locke Craig's urging, gave the Corporation Commission the power to set rates but did not require that it do so. The Raleigh *News and Observer*, owned and edited by Josephus Daniels—ardent Democratic champion of reform (including, oddly enough, the disfranchisement of North Carolina's black voters) and inveterate foe of the Republican Dukes and the "tobacco trust"—applauded the legislature's action. The newspaper had earlier warned that, "We are in danger of a power and lighting trust in North Carolina and it ought to be regulated before it has a chance to exploit the people."[33]

As the idea of *deregulating* the electric utility industry gained momentum and, in some states, began to become a reality in the 1990s, a distinguished political economist succinctly stated the original understanding about the regulatory system that prevailed throughout much of the century: "For most of the century the electric utility industry operated under an arrangement that gave the local utility companies regional monopolies and the promise of a fair opportunity to recover their prudently incurred costs in exchange for regulatory limitation of their profits and their assumption of an obligation to make whatever investments are required to ensure that the power comes on whenever anyone flicks the switch."[34]

The careful reader should note that the above sentence is complex and somewhat like a juggler with, say, a half dozen balls in the air. And the simple fact was that in 1913, when the whole idea was new in North Carolina, none of the parties involved—the Corporation Commission, the electric utilities, and the public at large—fully understood the rules or knew how to play the new game.

At any rate, the Corporation Commission in early 1914 ordered the electric utility companies in the state to submit their rate schedules and corporate records at the end of each year. Southern Power complied with the commission's order but only "in deference to the request...(and) not because any legal obligation is imposed...." The company went on to declare that it possessed the authority to set its own rates: "Each [rate] case must be treated on its peculiar circumstances, and the rates are subject to the reasonable rules and regulations of the company."[35] Since the Corporation Commission had received no particular complaint or case concerning Southern Power's rates, the Commission tacitly acquiesced in Southern Power's position, and there the matter stood for several years.

Because of inflation after World War I and other factors, Southern Power in 1919 set out to increase its revenue by raising some of its rates. The company had for a number of years sold bulk or wholesale electricity to the North Carolina Public Service Company of Greensboro at a rate of eleven mills (1.1 cents) per kilowatt hour. (A mill equals 1⁄10 of a cent.) The Public Service Company, in turn, retailed the power in Salisbury and other nearby towns. With the contract between Southern Power and the Public Service Company expiring in August 1918, Southern Power proposed a new contract for a period not less than five years, initially at a rate of fifteen mills (1.5 cents).

When the Public Service Company refused to sign the new contract, Southern Power gave a year's notice for the termination of service. Thereupon the Public Service Company obtained an order (mandamus) from the Superior Court of Guilford County compelling Southern Power to continue service. The Public Service Company's suit against Southern Power was based largely on the fact that Southern Power had a long-term contract (until 1944) to sell electricity to its own subsidiary, the Southern Public Utilities Company, at a rate of eleven mills per kilowatt hour but now proposed to charge the Public Service Company fifteen mills (subsequently increased to 18.8 mills). Arguing that a rate increase was "absolutely necessary to earn any income," Southern Power appealed the Superior Court's order to the North Carolina Supreme Court.[36]

Late in 1919, in a split three-to-two vote that went against the Southern Power Company, the state's Supreme Court rendered a decision that the *Charlotte Observer* described as one of the most important of the previous hundred years. In writing the majority opinion, Chief Justice Walter Clark had a field day, for not only was he widely regarded by much of the public (and by himself) as the foremost liberal in the state, but he had waged fierce battles against the Dukes and their interests since around the turn of the century. Despite some impassioned and extreme passages in his long opinion, Clark, in a calmer mode, made this quite reasonable assertion: "The defendant power company is a public service corporation and by reason thereof enjoys the right of eminent domain under which it has been enabled to construct and operate these [transmission] lines.... It enjoys therefore a monopoly of this business, and also by long service its business is 'affected with a public use' and it is therefore subject to public control and regulation not only in fixing and prescribing its rates, but more especially in the requirement that it shall furnish its facilities at the same rate to all receiving them under like conditions." Then in a telling allusion to some then-recent history, Clark declared that the Supreme Court's object was not to fix or judge rates but to prevent discrimination between the purchasers of Southern Power's electricity, "which is a method by which the Standard Oil Company, the American Tobacco Company and all other trusts have crushed opposition and enlarged their power and increased their accumulations to a point which made them a menace to governments by the people, and caused their dissolution by judicial decree."[37]

Although the Southern Power Company appealed to and ultimately failed in the federal courts, it acquiesced in the North Carolina Supreme Court's decision and filed an application for a rate increase with the Corporation Commission late in 1920. The importance of the Commission's hearing on the matter was underscored not only by the crowded courtroom but also by the presence of J. B. Duke, making a rare appearance in Raleigh, along with W. S. Lee, Norman Cocke, and others. In presenting the power company's case, attorney Zebulon V. Taylor declared that up to that point the company had invested almost $47 million in North Carolina alone. The earnings in the previous year, however, had not aggregated 3.81 percent, and during the previous five years the earnings of the company had not reached the 5 percent mark. The time had come, Taylor concluded, for the company to show better returns on its investment if it wished to expand and gain additional capital.[38]

With the Corporation Commission calling for a continuance of the case until mid-January 1921, the action now shifted to the state legislature—and a classic example of how politics interacts with and often distorts rate regulation. Josephus Daniels and his Democratic allies in the legislature had hailed Judge Clark's decision and the principle of state regulation of electric utilities. Under the new circumstances, however, they clamored for—and almost got—a state law that would leave all existing contracts between Southern Power and its large customers (mostly textile mills) outside of any rate adjustments that the Corporation Commission might be willing to make for Southern Power.

While the owners of some cotton mills recognized the seriousness of the power company's plight and its need for higher rates, others including the Cannon Mills, did not and fought tenaciously to protect the low rates in their existing contracts. In a public hearing before the state senate's committee on the Corporation Commission, the attorney for the mills (that is, those that supported passage of a law that would protect their existing contracts) argued that the owners of the mills had entered into contracts with the power company in good faith and had scrapped their former equipment because they were offered electric power at an attractive price. Further, he declared that the legislature of 1913, which extended the powers of the Corporation Commission, had never dreamed that the power would be used to abrogate contracts and that the legislature should withdraw the power that had been given inadvertently by its predecessor and had never been exercised.[39]

W. S. O'B. Robinson, principal attorney for the power company, countered that the company also had entered into contracts with good faith and had not sought to be relieved of them until the recent decision by the state's Supreme Court. That decision, Robinson noted, held that all rates of the power company were subject to review by the Corporation Commission and that contracts would not hold as against rates fixed by the Commission.

Other lawyers as well as W. S. Lee also spoke, but, as a newspaper reporter observed, J. B. Duke sat quietly during the discussion. Only after the hearing ended did he remark, "I think our boys made the best speeches."[40]

The legislative battle raged for many weeks, and the *News and Observer* and *Charlotte Observer* took diametrically opposite positions. The former wielded its greatest influence and power in the still-predominantly agricultural eastern portion of the state, while the latter championed the interests of the rapidly industrializing Piedmont. "There will be little sympathy for the demands of the power companies to be allowed to make larger profits,"

Daniels' paper declared. "The emergency demands that in behalf of the people the grave responsibility which a previous Legislature entrusted to the Corporation Commission be recalled to the extent that it may under no circumstances be employed to permit the annulling of existing contracts. . . ."[41]

The *Charlotte Observer*, on the other hand, hoped that the public would soon better understand "the influences of prejudices and animosity" that the Southern Power Company had to live down before it could attend to its "program of continued development work in the waterpower and interurban [railway] territory of this part of the South." The Charlotte paper avowed that it saw the matter "not as a fight on Duke and on his power company, but as a fight against the industrial interests of this part of North Carolina." It was a situation in which "Industrial North Carolina should at least match interest with Political North Carolina."[42]

After an unusually intense political battle, the measure to force Southern Power to maintain its existing contracts with textile mills passed the state senate but failed by one vote in the lower house. The *Charlotte Observer* rejoiced in the outcome of what it termed one of the greatest fights in the legislative history of North Carolina. "It is a very easy matter for legislation to tie the hands of capital," the Charlotte paper noted. Yet "no faction or combination of factions, however animated, whether by personal prejudice or political animosities, could strike [J. B.] Duke to the ground and not kill the most momentous factor that has yet been created in this part of the country in [the] development of textile and related industries, and in promotion of the prosperous conditions which follow the establishment and operation of these industries." Another happy circumstance, the *Charlotte Observer* believed was "the establishment by the Legislature of a righteous vindication for the Corporation Commission."[43]

The *News and Observer* saw the matter differently, of course, and as the action shifted back to the Corporation Commission and its hearing on Southern Power's application, Josephus Daniels let loose a steady barrage of fiery editorials. "The people of North Carolina are threatened with a monopoly in power, in lights, in electric transportation, in gas by reason of the methods of the Southern Power Company," the Raleigh editor asserted. "The Corporation Commission must protect the people or Monopoly and Extortion will have them by the throat."[44]

Despite such rhetoric, the Corporation Commission in July 1921, granted at least a part of Southern Power's request. Hoping to achieve a return on investment of 6.86 percent, the company asked for a wholesale price of 1.4 cents per kilowatt hour. The Corporation Commission, however, granted a rate of only 1.25 cents, subject to change upon any subsequent petition and additional data from the power company. Although no doubt glad to win even a partial victory, J. B. Duke and his associates in the power company were determined to come knocking again on the door of the Corporation Commission.

One reason, among many, that J. B. Duke wished to increase the earnings of the power company was that until that could be accomplished, he had his plans for large-scale philanthropy in the Carolinas on hold. Since the charitable trust he envisioned would be based substantially on a large portion of his stock in the power company, he believed that he had to wait until the annual income from the stock would be sufficient to accomplish his goals.

Beginning around 1916, J. B. Duke had begun to intimate to the president of Trinity College, William P. Few, that he (Duke) planned to do something big for the college. In 1919 he went further and revealed to Few some of his thinking about linking his philanthropy with the power company; at J. B. Duke's request, Few, although unwell at the time, accompanied Duke to one of the important hearings in Raleigh in the spring of 1921.

About that same time Few submitted to Duke a memorandum in which he outlined a plan to organize a major research university around Trinity College. There was already a Trinity University in San Antonio, Texas (and numerous other collegiate institutions named Trinity scattered over the English-speaking world). Moreover, Washington Duke had played the central role in bringing Trinity to Durham in 1892, and he and his family had gradually become the principal, indispensable backers of the college. President Few suggested, therefore, that the new university should have an appropriate and distinctive name—Duke University. While J. B. Duke was simply not ready to act in 1921, for the reason explained above, he gave Few strong reasons to hope.[45]

President Few had no choice but to wait, yet the demand for electric power in the Piedmont Carolinas would not wait. New generating plants, requiring new capital, had to be built in a steady procession. Including the new stations at Bridgewater and Wateree, Southern Power had eight hydroelectric plants (seven of them on the Catawba) and four auxiliary steam plants by 1920. In the aftermath of the 1921 rate increase, limited though it was, J. B. Duke authorized the construction of two new hydro plants on the Catawba, Dearborn at Great Falls, South Carolina, with a generating capacity of 54,000 horsepower, and Mountain Island near Mount Holly, North Carolina, with an installed capacity of 80,000 horsepower. Moreover, the company added units to the existing steam plants at Mount Holly and Eno Station (near Durham).

With all that completed or approaching completion, however, J. B. Duke finally blew the whistle: he announced that if the power company could not obtain what he and his associates considered a fair return on investment, then there would be no more plants built in the Piedmont. With the company applying to the Corporate Commission for an increase in its wholesale rate to 1.4 cents per kilowatt hour (the amount requested but denied in 1920–1921), J. B. Duke issued a rare public statement in October 1923. "I am ready to proceed to spend more money, to build more plants to create more power for the future development of the Carolinas," Duke explained, "but I am not willing to spend it on the basis of the returns which the Southern Power Company is now allowed." He noted that the company had averaged a return of not more than 4 percent since it was started, and those small earnings had always been plowed right back into the company, along with additional millions.

"I have put approximately $60,000,000 of my own money into the Southern Power Company so far," Duke continued. "I have never taken a cent out of it, and never expect to." What he was to invest in the company in the future depended on the attitude of the people toward the undertaking, as revealed through their representatives in Raleigh, the three members of the Corporation Commission.

"We will lay the situation before the Corporation Commission," Duke stated. "Our books will be submitted to that body. . . . We will let the records speak for themselves." If the Commission granted the requested increase, then the company was prepared to build another plant costing $10 million to supply the already existing demand for electric power. "Otherwise, I am through," he concluded.[46]

Tom-toms reverberating, Josephus Daniels took to the warpath. The steady drumbeat of flaming editorials in the *News and Observer* began in October 1923, and charged, among other things, that J. B. Duke, a die-hard monopolist from way back, was out to monopolize electric power in the Carolinas; that much of the so-called investment in the power company was actually "watered stock"; and that since J. B. Duke thought he was either Superman or King, the Corporation Commission might as well abdicate.[47]

The *Durham Herald* somewhat meekly observed that "it is a pity that the *News and Observer* is not willing to allow the Corporation Commission to pass upon the [rate] case in accordance with the facts." To which the Raleigh paper replied that it merely wanted the Commission to consider all the facts, including "the use Mr. Duke has made of monopoly in the past and to determine whether, in the light of the record, he ought to be confirmed in monopoly of so vital a things as power."[48]

As influential as Josephus Daniels was in Democratic circles, there were others, besides the *Charlotte Observer*, who dared to take him on. W. O. Saunders, a salty newspaperman in Elizabeth City in the northeastern corner of North Carolina, called on J. B. Duke in his New York office. Saunders asked when Duke was going to bring his power lines into eastern North Carolina, which lagged glaringly behind the Piedmont, and "give us coast country folks some of the cheap power that his great superpower system is feeding to scores of towns and cities in the Piedmont section today."

Duke replied that there was indeed power potential aplenty on the Roanoke River in eastern North Carolina. He thought, however, that Josephus Daniels should develop it. Daniels, Duke continued, is a "big, constructive man, always wanting to do something for the people." Let Daniels build a monument to himself, Duke suggested, by harnessing the Roanoke and "putting it to work for eastern North Carolina, just as the Southern Power Company has put the mountain streams of the Carolinas to work for the west."

If there was any bitterness in Duke's words, Saunders reported, he did not betray it in any look or tone—but "there was a sly twinkle in his eye." In the meantime, the Elizabeth City editor noted, Duke was spending millions on hydroelectric projects in Canada. "But Mr. Duke meets with less opposition in Canada than in his native North Carolina. Canadians seem glad enough to let him take their waste waters and turn them into power to be sent over the border to run New England mills. But in North Carolina he meets the constant opposition of Mr. Daniels and the *News and Observer*."[49]

A cotton mill owner in Gastonia, who, like quite a few other mill owners in the Piedmont, supported the power company's request for a rate hike, found another way to tweak the nose of Josephus Daniels. The Gastonian pointed out that Daniel's opposition to the requested rate hike seemed odd in light of the fact that since 1917 the *News and Observer* had raised its advertising rates approximately 60 percent and its subscription rates nearly $3.50. Would the *News and Observer* be willing to go back to 1917 rates?[50]

Taking a more serious approach, *Natural Resources*, a publication of the State Geological and Economic Survey, declared that the greatest danger to the further development of North Carolina's water power lay in the fact that "the politician has added the word 'kilowatt' to his vocabulary." The publication reported that North Carolina then ranked fifth among the states in water power development and second only to New York in the states east of the Mississippi. The forecast was that North Carolina would need an additional 120,000 horsepower in the next two years and an additional 469,000 horsepower for the next five years. At a conservative cost estimate of $100 per horsepower, that meant a capital investment of $12 million in two years to be followed by $46 million in the five following years. In other words, there needed to be made available for North Carolina in seven years more horsepower than had been installed in the whole state up to that point. *Natural Resources* next asserted that the Southern Power Company had not only played the leading role in the past but "its activities constitute[d] the best hope of the immediate future." *Natural Resources* concluded that there was nothing in the big, pending rate case that could be brought into politics "except through sheer recklessness, demogogism or wickedness."[51]

As the war of words raged on in the newspapers, W. S. Lee and his associates sparred with attorneys representing Cannon Mills and a few others before the Corporation Commission. (J. B. Duke stayed away from Raleigh for this round.) Lee presented an array of exhibits purporting to show that on the basis of any one of several valuations of the power company's property, including the one used by the Corporation Commission in 1921 plus actual investments made since then, the return to the company under the proposed rate hike would be less than 8 per cent. One of the attorneys for the company argued that not a man in the cotton mill business would undertake to keep his mill operating if he were limited to an 8 percent return on investment.

Lee testified that in North Carolina in 1922 power sales had increased almost 22 percent over 1921 sales; in 1923, the increase was almost 14 percent over 1922; and the estimate was a 24 percent increase for 1924. (There had been much less increase in South Carolina sales for those years, Lee interjected.) "If North Carolina goes ahead as it reasonably should," Lee declared, "we've got to build one or two power plants a year to keep up with the business or somebody else has to do it."

As to average earnings on invested capital in the Piedmont territory served by the company, Lee noted: "I don't think there are any [companies] but what are earning 12 or 15 percent on the invested capital and some of them are going much higher than that." He reminded the Commission that Southern Power had employed an independent engineering consultant to evaluate its property. Moreover, in the 1920–1921 hearing, the protestors (later to be called intervenors) had called for an outside investigation of the power company's books and records. That request was granted, and those then opposed to the rate hike had employed accountants to check the company's books and engineers to inspect the physical properties. After that had been done, no word of testimony was heard from the protestors. Under the rate approved in 1921, the company maintained and the Corporation Commission agreed that the return was 5.91 percent on investment. Under the rate proposed in 1923–1924, there was similar agreement that the return would be 6.86 percent.[52]

J. B. Duke and Southern Power won the desired increase in January 1924. In its long order approving the new rate, the Corporation Commission declared, among other things, that the power company was undertaking in good faith to supply a service for which there was an ever-increasing demand. Within the previous two years it had added 200,000 horsepower to its system, and those additions alone constituted a greater volume of electricity than was produced by all the other companies in North Carolina combined. Those new power units were capable of turning the wheels of more than four hundred cotton mills of ten thousand spindles each. The company was "entitled to a fair return for such service," the Corporation Commission concluded, "and such rate of return as will justify the continuing investment of such sums of capital as may be necessary to continue to meet the increasing demand for power for new industries in the territory it serves."[53]

The power company had finally won a crucial battle and would not again face such a critical struggle about its rates until the difficult, indeed perilous, 1970s. It was not as if J. B. Duke and others in his family and those allied with him were out to make a pile of money. Indeed, J. B. Duke once confessed to his friend Mrs. E. C. Marshall that he had come to realize, just when he did not say, that there simply was not the money to be made in the electric utility business that there had been in tobacco. Yet money-making, per se, was not Duke's prime purpose in his zealous effort to build up and protect the power company. Encouraging the industrialization of the Piedmont Carolinas and providing a stable support-base for his projected charitable trust were his true purposes.

Things were coming together nicely for J. B. Duke in 1924. Not only had the Southern Power Company been given a green light to go ahead, but Duke, after almost a decade of effort, had cleared all obstacles to his mammoth Canadian project. He was also on the way to working out an advantageous disposition of his immense investments on the Saguenay.

He would announce the creation of the Duke Endowment in early December 1924, and the establishment of Duke University would be announced later in that same month.

An unprecedentedly dry summer in 1925 would pose a formidable problem for the power company. While J. B. Duke finally approved a long-range solution to avoid such a problem in the future, death would take him by surprise in October 1925.

~

As J. B. Duke had promised, the Corporation Commission's approval of the rate increase in early 1924 was the green light for the power company to launch a large expansion. With demand for electricity, particularly for textile mills, steadily mounting in the Piedmont, the company scrambled to stay ahead. Three new hydro-stations and one new steam plant were begun in 1924–1925, and, as will be discussed later, construction also began on an especially important steam plant late in 1925. It was destined to be the first large-scale coal-fired steam plant in the system and, although no one could realize it at the time, it would mark the beginning of the shift from hydro to steam, from the Catawba's free "white coal" to fossil fuel that had to be mined from the earth and imported some distance to the Piedmont.

The first new hydro station to be built was at Rhodiss, North Carolina, on the Catawba north of Charlotte. With an installed generating capacity of 42,400 horsepower (25,500 kilowatts), the plant went into operation in early 1926. Simultaneously with its construction, work began on a new auxiliary steam plant, named Tiger, near Duncan, South Carolina, in Spartanburg County, and before 1924 ended it was producing 30,000 kilowatts.

At what had been India Hook Shoals, where Dr. Wylie and W. S. Lee built the company's first power plant, a somewhat unusual arrangement resulted in a significant increase of productivity. The power house of the original Catawba plant was dismantled, filled with cement, and then covered by masonry to become part of a new spillway. The company's imaginative engineers then built a new powerhouse on top of the old, now-raised dam and located on the opposite side of the river. Whereas the original plant had a generating capacity of 9,400 horsepower, the New Catawba plant came in at 84,000 horsepower.[54] (In 1960 New Catawba was renamed Wylie Station, in honor of Dr. W. Gill Wylie, and the large reservoir renamed Lake Wylie.)

The last of this mid-decade group of plants to be built was Cedar Creek Hydro Station. Located on the Catawba at the Rocky Creek development in South Carolina, the new plant had a generating capacity of 57,000 horsepower and went into operation in August 1926. This brought the total of the new capacity of the four plants that were installed in the immediate aftermath of the rate increase to 223,400 horsepower—and the company had begun with a plant that produced only 9,400 horsepower in 1904.

That there was a ready market for all this electricity was a source of great satisfaction to the company's president, for more than one reason. Aside from wanting a fair return on investment, J. B. Duke had hoped to spur industrialization in the Piedmont Carolinas, and by the 1920s the statistics revealed that his power company was indeed helping to achieve that goal.

"One of the most persistent impulses in the life of the South since the Civil War," a prominent historian has declared, "has been the desire to develop an industrial economy."[55] Certainly a basically important precondition for the pursuit of that goal was the mania for cotton mills that swept over the region from the 1880s well into the twentieth century. Moreover, another well-known historian has described the shift of the cotton textile industry from New England to the South's Piedmont between 1880 and 1930 as "the most important migration ever experienced by a major branch of American industry...."[56]

Why did that shift occur? Discounting fervent southern hopes and prayers—and mountains of boosterish rhetoric—one finds that two of the most frequently offered and widely accepted explanations center around lower labor and transportation costs for the southern textile mills. While both of those explanations have undoubted validity, one historian has pointed out that if the Piedmont South had not developed an adequate and cheap source of power, cheap labor and lower transportation costs would have been meaningless. "Electrical power, generated and transmitted over long distances," according to this historian, "was a major factor in the shift of the cotton industry to regions previously associated with an agrarian economy."[57]

The argument continues by noting that New England textile manufacturers relied mostly on local generation of electricity at the mill sites. In the Piedmont Carolinas, on the other hand, the textile mills after about 1905 "rented" most of their power from a centralized system, with its generating plants located on the Catawba and a few other rivers of the Piedmont. Since "rented" power allowed for lower capital expenses, reduced operating costs, and increased flexibility in locating textile mills, the southern mills gained a distinct economic advantage over those of New England and thus guaranteed a continued migration of the cotton textile industry to the South.[58]

When the Southern Power Company was chartered in 1905, the cotton textile industry was well established in the Piedmont Carolinas, which already had over 20 percent of the nation's spindles—and half of those operated in a hundred-mile radius of Charlotte. Then, however, the typical southern mill got its power from steam engines. As mentioned earlier, J. B. Duke, W. S. Lee, and others in the power company set out to change that arrangement and achieved remarkable success fairly quickly.

As early as 1913, southern textile mills consumed more cotton than those of New England. Then in the 1920s the southern mills clearly achieved national supremacy: in value of product, 1923; in total spindles and total wages, 1927; and in looms between 1927 and 1931. Moreover, despite a popular misconception, the growth of the southern textile industry did not result from a wholesale migration of plants and capital from New England. A study in 1922 revealed that almost 84 percent of southern spindles were owned or controlled by southern capital.[59]

Unfortunately many historians and other writers have long tended to denigrate the cotton textile industry. Forgetting the rather common-sensical point that industrial economies must learn to walk before they can run, these historians bemoan the fast that the South did not immediately move into diversified manufacturing and high technology. Moreover, in countless books that emphasize the southern textile workers' low wages and long hours, the authors tend to ignore or gloss over the fact that the wages, as low as they were indeed, were still better than those to be made from tenant farming. And one has to wonder if the working hours of southern farmers, whether landowners or tenants, were any shorter than those of the mill workers.

The most influential historian of the late-nineteenth and early-twentieth-century South, C. Vann Woodward, in a chapter tellingly entitled "The Colonial South," paints a largely negative and gloomy portrait of the region's struggle for industrialization in that era. Admitting that there were gains in the South's cotton textile production, he nevertheless argues that the industry "did not escape the general pattern of colonialism [i.e., subordination to northern capital] to which other industries of the region were prone to conform." Noting that the chief products of the southern mills were yarn and coarse cloth, much of which got shipped north for final processing, he adds that the investment of northern capital and the "dependence upon Northern commission houses for selling and even for financing brought the cotton mills into line with the colonial trend."[60]

J. B. Duke did not, of course, have the pleasure of living to see Charlotte become one of the nation's preeminent banking centers. He would have quickly understood, however, that such could not have happened if the city had not first become a center for the financing of cotton textile manufacturing, a fact made possible at least in part by the power company that was so close to Duke's heart.

If the rate increase allowed in early 1924 only indirectly spurred more power plants and more textile mills, it operated in a more direct fashion to trigger J. B. Duke's long-contemplated Grand Design for perpetual philanthropy in the Carolinas.[61]

Work on Duke's philanthropic plan accelerated in the latter half of 1924, and in early December of that year he assembled a sizable group in his home in Charlotte to make a final assessment of the legal instrument, the indenture, that would establish the charitable trust. Among those in the group were his principal lawyer, W. R. Perkins, and his closest business associate, George G. Allen; and from the power company, Norman Cocke, Edward C. Marshall, and Charles I. Burkholder. The indenture had been drafted chiefly by Perkins, and for several days the group pored over it paragraph by paragraph. According to Norman Cocke's memory many years later, the Charlotte group suggested no substantial alterations or additions, and on December 9, 1924, newspapers across the nation carried the news of the establishment of the Duke Endowment, the South's largest philanthropy up to that time. (Duke signed the indenture at his legal residence in New Jersey on December 11, 1924.)

J. B. Duke placed securities, the great majority of which consisted of stock in the power company, worth approximately $40 million in his charitable trust. The fifteen self-perpetuating trustees of the Endowment were directed, first, to divide equally among themselves 3 percent of the annual income from the trust as compensation for their services. Then they were to set aside 20 percent of the remaining annual income to be added to the trust's corpus (principal) until an additional $40 million had been accumulated. (This was accomplished by 1976.)

Then the remaining 80 percent of the annual income was to be distributed in the two Carolinas as follows: 46 percent to four educational institutions (32 percent to Duke University, 5 percent to Davidson College, 5 percent to Furman University, and 4 percent to Johnson C. Smith University); 32 percent to help with indigent care and to help build nonprofit community hospitals for both whites and African Americans; 12 percent for certain specified purposes of the Methodist Church in North Carolina; and 10 percent for orphanages for both races.[62]

Concerning the power company, Duke had a good bit to say in his indenture. While much of the indenture consists of mind-boggling legalese and almost unintelligible (to lay persons) marathon sentences, there are certain scattered bits of clear and even affective prose, such as this:

> For many years I [J. B. Duke] have been engaged in the development of water powers in certain sections of the States of North Carolina and South Carolina. In my study of this subject I have observed how such utilization of a natural resource, which otherwise would run in waste to the sea and not remain and increase as a forest [does], both gives impetus to industrial life and provides a safe and enduring investment for capital. My ambition is that the revenues of such development

> shall administer to the social welfare, as the operation of such developments is administering to the economic welfare, of the communities which they serve. With these views in mind I recommend the securities of the Southern Power System (the Duke Power Company and its subsidiary companies) as the prime investment for the funds of this trust; and I advise the trustees that they do not change any such investment except in response to the most urgent and extraordinary necessity; and I request the trustees to see to it that at all times these [power] companies be managed and operated by the men best qualified for such a service.[63]

Not content with merely "advising" the trustees on the matter of investing in the power company, elsewhere in the indenture Duke stipulated that the trustees could not sell any of the Endowment's stock in the Duke Power Company except by the unanimous vote of all fifteen trustees at a specially called meeting. (This requirement would pose a sticky problem for the trustees later in the century.) Furthermore, the trustees were empowered to lend Endowment funds only to the power company and to invest any of the Endowment's funds only in Duke Power securities or certain carefully described categories of government bonds.[64]

When J. B. Duke made these arrangements to link the Duke Endowment so closely with the Duke Power Company, no one questioned the propriety of thus tying together a charitable trust and an investor-owned, profit-seeking business. Many other foundations, such as Kellogg and Ford, had similar links, although not necessarily with all the limitations that J. B. Duke had set. By the 1960s, however, public policy views would change, and the Tax Reform Act of 1969 would force significant modifications of J. B. Duke's Grand Design.

Not only had Duke attempted to make sure that the Endowment would always be there to help Duke Power meet its constant need for capital. He had also, in effect, given the virtual ownership and control of Duke Power to the Duke Endowment, as represented by its trustees. When J. B. Duke stated in the indenture that he wanted "the revenues of such [water power] development" to administer to the social welfare of Carolinians, however, he might more accurately have said "most of the revenues."

The qualification, however slight, would have accommodated the fact that he did not give all of his Duke Power stock to the Endowment. At the same time that he established the Duke Endowment, he set up the Doris Duke Trust. Two-thirds of the annual income of this private, family trust was to go to Doris Duke and one-third to J. B. Duke's nieces, nephews, and their lineal descendants. Although Duke placed only $35,000 in cash and

2,000 shares of Duke Power stock in the trust at the time of its creation, by his will he added 125,904 shares of the stock. Temporarily this meant that the Doris Duke Trust actually held more shares of Duke Power stock than did the Endowment. As J. B. Duke intended, however, that situation quickly changed by virtue of the restrictions on the Endowment's investments, and the Endowment's Duke Power holdings soon dwarfed those of the Doris Duke Trust. In 1968, for example, the Endowment owned over 56 percent of Duke Power's common stock and the Doris Duke Trust owned only a bit less than 9 percent. Since all of the trustees of the Endowment (except Doris Duke, who was to become a trustee of the Endowment upon reaching the age of twenty-one in 1933) also served as trustees of the Doris Duke Trust, it is clear that J. B. Duke had carefully made interlocking arrangements concerning the Duke Power Company, the Duke Endowment, and the Doris Duke Trust.

At any rate, J. B. Duke had a number of reasons to look on life with satisfaction as 1924 ended. The power company was receiving a fair return on investment and expanding significantly. The Endowment had been established for perpetual philanthropy in the Carolinas, and he had made additional provisions for his only child and a generous arrangement for his close relatives. There was to be a huge building program at Duke University where the old Trinity Campus (soon to be known as the East Campus) was to be redesigned and equipped with eleven handsome buildings in the Georgian or Neo-Classical style of architecture. On the new West Campus, a vast series of Tudor Gothic buildings constructed of colorful native stone would be erected once the new buildings on the East Campus were approaching completion (1927). In addition, Duke University became the owner of around 5,000 acres (later increased to about 8,000 acres) of what would become the Duke Forest. All of this, at a cost of around $19 million, was a gift to Duke University from J. B. Duke. (Later the Endowment came up with an additional $2 million for the soaring Gothic-style Chapel that J. B. Duke had wanted at the center of the West Campus.)

J. B. Duke enjoyed construction work, especially that which involved stone, so after taking a leading part in architectural and landscaping decisions concerning Duke University, he looked forward to the unprecedented building program that was about to get underway in Durham.

As if all that were not enough, there was also a happy denouement to his Canadian venture. In 1922 he and his associates had finally secured the permission of the provincial government of Quebec to raise the level of Lake St. John. That was the signal for J. B. Duke to put W. S. Lee, Frank Cothran, Richard Pfaehler, and other Duke Power engineers back to work on the design for what would be one of the world's largest hydroelectric plants (at that time) at Isle Maligne on the Saguenay, the upper power site on the river.

Bringing two decades of experience in building dams and hydro plants on the Catawba, these able engineers faced a new challenge on the Saguenay. Its natural flow was six to seven times that of the Catawba, and its energy potential much exceeded any existing demand in the region. The Isle Maligne plant's projected output, as finally worked out by Lee and his fellow engineers, would be 540,000 horsepower, whereas Southern Power's eight hydro plants in 1920 had a combined capacity of some 314,000 horsepower.[65]

Gambling on his ability to find and sign up consumers for the Saguenay power, as he had gambled to a certain extent on the Catawba, J. B. Duke late in 1922 joined forces with Sir William Price, an important figure in the Canadian newsprint industry. Price agreed to take for his paper company a large block of the electricity to be produced at Isle Maligne. He also took a minority interest in the Duke-Price Power Company, which built in 1923–24 the Isle Maligne dam and power plant.[66]

Since the manufacture of aluminum involved a process requiring abundant electricity as well as bauxite, J. B. Duke had engaged, off and on, in a world-wide search for an available source of bauxite after about 1908. The details of his search need not be recounted here, but in the early 1920s the prospect of having huge amounts of unsold electricity at Isle Maligne gave a certain urgency to the matter.

Adding immensely to the advantageousness of the Saguenay power for industry was the fact that in the relatively short stretch of about thirty miles below Lake St. John, the Saguenay completed its 300-foot fall to sea level. Then in the remaining fifty-two miles to the massive St. Lawrence River, the Saguenay ran through a broad, deep fjord that was navigable by ocean-going ships.[67]

After intensified efforts to find and gain control of a large source of bauxite and various other schemes—including the possibility of selling Saguenay power to New England utilities—J. B. Duke late in 1924 began a series of conversations with Arthur Vining Davis, head of the Aluminum Company of America. By April 1925, those negotiations resulted in J. B. Duke's exchanging his as-yet-undeveloped power site at Chute-à-Caron, the potential of which was even greater than that of Isle Maligne, for a one-ninth interest (about $17 million) in a reorganized Aluminum Company of America. Furthermore, Davis agreed for Alcoa to purchase 100,000 horsepower from the Isle Maligne plant.[68]

Gambling for enormous stakes on the Saguenay, J. B. Duke had arrived at a solution that he liked by the spring of 1925. He had put much capital into the Canadian venture, but in light of the outcome, the evidence suggests that his deepest interests and concerns remained focused on the Piedmont Carolinas. After a visit to Durham where the final decision was made concerning the location of Duke University's Tudor Gothic buildings, Duke later visited Charlotte to confer with his associates there and to tour various plants. Because his wife and daughter were at the family's "cottage" in Newport, Rhode Island, Duke felt lonely one evening and invited Norman Cocke to come over for a talk.

Cocke later recalled that Duke looked well but seemed in a somewhat reminiscent mood and talked about the early tobacco business, the power company's abortive effort to make money by farming on some of its lands near Great Falls, and various other topics. The Piedmont area was having a prolonged dry spell, but nothing critical by the time of Duke's visit.[69]

Vigorous and apparently in good health at age sixty-eight, J. B. Duke, together with George Allen, W. S. Lee, and various officials of the Aluminum Company of America visited the Saguenay in July 1925. Duke and his party travelled in his private railway car, the "Doris," and Arthur V. Davis, Secretary of the Treasury Andrew Mellon, his brother R. B. Mellon, Roy A. Hunt, who later became president of Alcoa, and others went on another private car.

Having already lost some weight at the doctor's instructions, Duke was still dieting and, according to Roy Hunt, grumbled about having only "three damned prunes" for breakfast. Duke carried a walking cane, as he had for some years, but so did Lee and Allen. Duke impressed Roy Hunt as alert and possessed of a sharp mind. Physically he kept up with the group in several days of fairly strenuous moving about in what was then a relatively undeveloped region. Altogether, Hunt sized Duke up, as has been mentioned earlier, as "soft spoken," a "courteous Southern gentleman" who seemed "very friendly and very keen."[70]

Having touched base at Duke University, Charlotte, and on the Saguenay, Duke joined his family in the latter part of July 1925, at Newport. There he became ill, and though Mrs. Duke continued to be hopeful about him through August, when his condition only worsened, he was carried in his private railway car to New York. There his doctors discovered in September that he suffered from what they diagnosed as pernicious anemia, a disease for which there was then no known cure.

As Duke lay dying in his Fifth Avenue mansion, the prolonged dry spell in the Piedmont Carolinas had turned into a record-breaking drought. Even before Duke left Newport, W. S. Lee and Charles I. Burkholder had gone up to confer about the power company's drought-caused crisis. Duke authorized them to announce that the company proposed "to put the system in shape to meet any possible contingencies" arising from a prolonged lack of rain.[71]

The power company had had earlier experience with trouble-causing dry spells. In 1922, for example, drought in the Carolinas produced potential power shortages and some temporary, rotating curtailments of service. Ironically, in one of the issues of the *News and Observer* that carried an editorial lambasting J. B. Duke upside down and backwards, the newspaper reprinted a story from the *New York Times*. Entitled "Waterpower and Statesmanship," the *Times* story hailed the "super-power system" that had been created in the Southeast through the linking together of seven independent but cooperating electric utilities. Buying surplus power from a generating plant at Muscle Shoals on the Tennessee River (a plant built during World War I and owned by the Federal government), an Alabama utility was thus enabled to sell power to one in Georgia. It in turn, through links that went back to 1914, could sell surplus power to the companies in the Carolinas that were facing power shortages because of drought.[72]

J. B. Duke, Lee, and other officers in the power company, had worried about the need for additional coal-burning steam plants even before 1925's drought. In fact, as mentioned above, one had been built in South Carolina in 1924, Tiger Plant near Spartanburg. Mrs. Duke told her old friend, Mrs. Marshall, that for two years her husband had "talked of nothing but steam plants," a subject that Mrs. Duke claimed not to understand.[73] Nothing in the way of earlier "dry spells," however, had quite prepared the company for the challenges it faced when the rains refused to come, week after week and then month after month, in the summer of 1925. The linkages with the "super-power system" were not enough this time, and from late August, 1925, to early 1926, the company had to curtail service materially.[74]

At one point during his final illness, J. B. Duke's private nurse noted that he seemed restless. When she inquired if he wanted anything, he responded, "Please don't disturb me. I'm building a steam plant down South."[75] Sure enough, after studying blueprints and estimates carefully prepared by Lee, Nabow, and other of the company's engineers, Duke had Norman Cocke and E. C. Marshall come up to visit him in his sickroom. There he gave final authorization for them to proceed with the construction of a large-scale steam plant on the Yadkin River near Salisbury, North Carolina. With a proposed initial capacity of 70,000 kilowatts, it would be the first of its kind in the Duke Power system and the first power plant in the South (and one of the first in the nation) to utilize new technology and equipment that allowed the burning of coal pulverized to a talcum-powder fineness. And though the fact may not have been fully realized at the time, the plant would become a significant benchmark in the transition of the Duke Power system from a primary dependence on water as the principal source of its power to steam, whether produced by burning coal or, at a much later date, nuclear energy.

J. B. Duke's approval of the large-scale steam plant evidenced his flexibility and willingness to adapt to new circumstances, even as he lay dying. The decision was undoubtedly not an easy one, however, for he had both economic and esthetic reasons to prefer "white coal." Less than a year earlier, he had spoken eloquently in his indenture creating the Endowment about the Carolina rivers as natural resources running "in waste to the sea." He had, he stated, wanted the revenues from the power developments on those rivers "to administer to the social welfare, as the operations of such developments is administering to the economic welfare, of the communities which they serve." There was sad irony in the fact that Duke felt he had, at the least, to begin abandoning one of the underlying premises of the indenture creating the Endowment even as it was just getting underway.

J. B. Duke died in his Fifth Avenue mansion on October 10, 1925. He would have been sixty-nine years old on December 23. After a simple, private service without eulogy or sermon in his New York residence, his body was carried by train to Durham. There, after services in Duke Memorial Methodist Church, he was buried beside his father in the family mausoleum in Durham's Maplewood Cemetery. Later, in the 1930s, the remains of Washington Duke and his sons, Benjamin Newton Duke and James Buchanan Duke, were placed in three marble sarcophagi in the small Memorial Chapel to the left of the chancel in the Duke University Chapel.

As for the Duke Power Company, J. B. Duke had built wisely and well. George G. Allen became and long remained its president; W. S. Lee continued as chief engineer and vice-president; and the company, blessed with high morale and a deep awareness of its singular history, prospered along with the area it served in the late 1920s.

It is sometimes said that, in the last analysis, Duke University stands as the most visible memorial to J. B. Duke. Perhaps. As concerns the economic well being of several millions of Piedmont Carolinians through most of the twentieth century, however, the Duke Power Company is itself also a remarkable monument to J. B. Duke, Tar Heel entrepreneur and capitalist.

Introducing a moral judgment into the dynamics of capitalism, one commentator has made this intriguing observation: "Genuine entrepreneurs are impelled by a deep-seated urge to create something of value."[76] If that be so, as one must suspect that it is granted the qualifying word "genuine," then in the Duke Power Company J. B. Duke had indeed created "something of value."

Dr. W. Gill Wylie

William States Lee (I)

James Buchanan Duke
as a young man

Benjamin Newton Duke

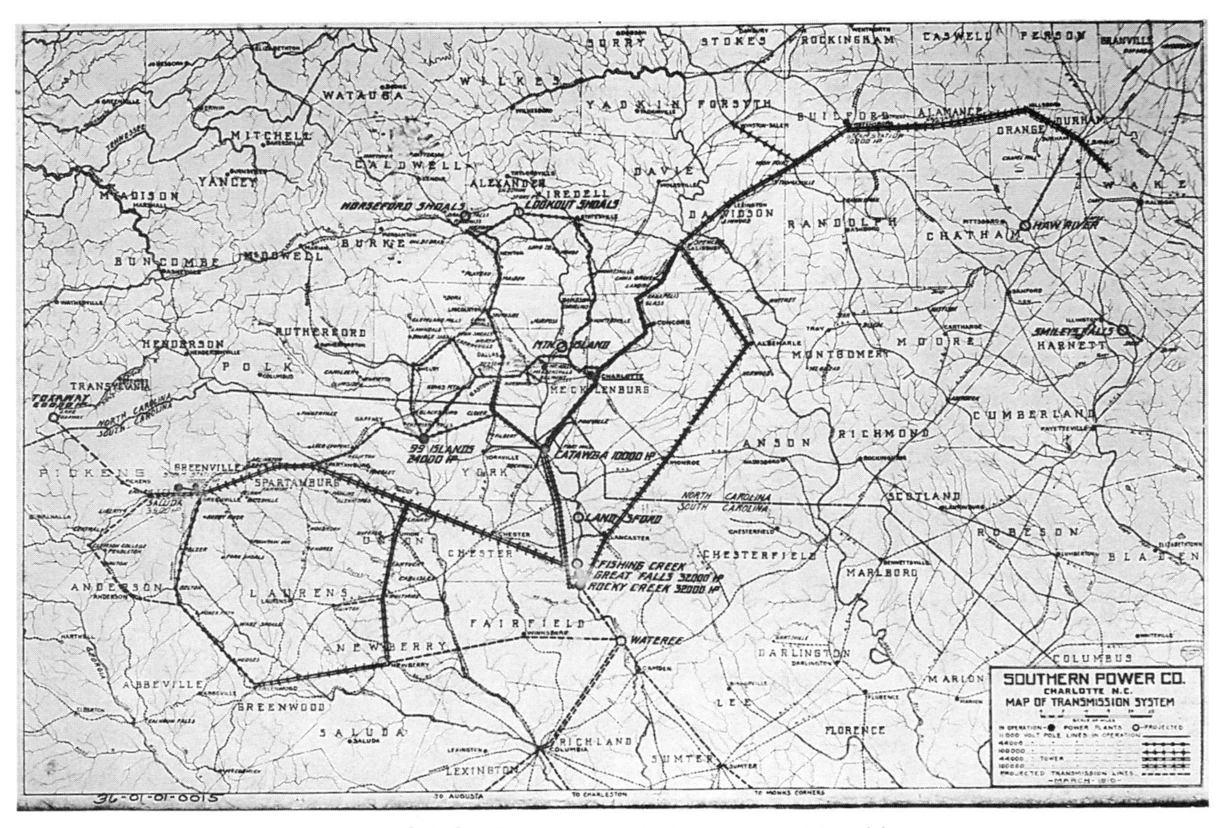

Map of Duke Power transmission system, 1915 (?).

Dedication of dam at Great Falls in 1906. Front row, l. to r.: W. S. Lee, B. N. Duke, Dr. Gill Wylie, J. B. Duke, and Mary Lillian Duke.

George Garland Allen

Norman Cocke

Edward C. Marshall

Charles I. Burkholder

Construction camp workers all dressed up in 1910.

Construction camp workers and tents.

Linemen set power pole.

Workers showing off on transmission tower.

Transformer being moved by horse and wagon.

Construction camp musicians dressed in Sunday clothes.

Home-made pole trailer.

Moving a transformer with oxen.

Two speedy servicemen.

Record-breaking flood on the Catawba River, 1916.

Boating on the Great Falls reservoir.

Rhodiss Hydro Station with a textile mill on right across river.

A worker stands beside giant turbine at Bridgewater, 1918.

Duke Power Building under construction in Charlotte, 1928.

Buck Steam Station under construction, 1926.

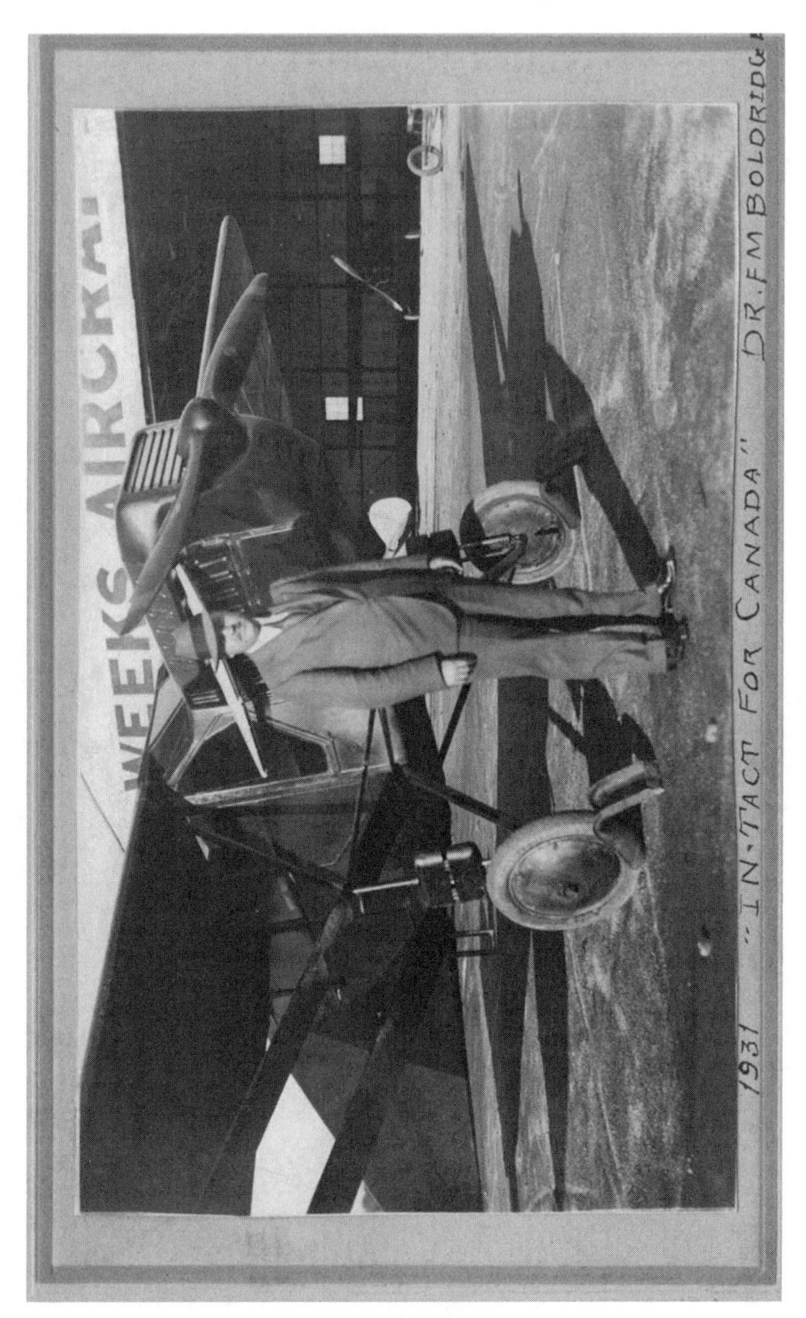

Dr. Frank M. Boldridge, 1931.

Promoting appliances in Greenville, South Carolina, early 1930s (?).

Two Duke Power home economists demonstrate the "latest things" in early 1930s.

An appliance promotion in Charlotte.

CHAPTER III

FROM THE ROARING 'TWENTIES TO THE DEPRESSED 'THIRTIES AND WORLD WAR II

In 1924, the year before J. B. Duke's death, the power company decided to embark on a new policy. It was one that would further set it apart from most other electric utilities in the nation and would play a major role in its ability to produce electricity more efficiently and economically than could many other utilities. The company's outstanding cadre of engineers had from the first done the designing of both its hydro and its steam plants. Now, however, from 1924 forward the power company would do its own construction work. This "do-it-yourself" policy quickly became an immense source of pride and enhanced morale for Duke Power as well as a significant source of cost-reduction.

The origins of the new policy were, at least partly, to be found in legal difficulties that arose between the company and certain contractors following the flood of 1916. The practice had been for the company to clear the trees from areas that were to be inundated by a new reservoir or to contract for that to be done. At the time of the sudden flood in 1916, however, there were many trees that had been cut down but not yet removed or burned in certain areas. As the Catawba rose, it picked up the debris, together with everything else coming downstream, and carried it along with the flood waters.

According to later reports of "old timers," the debris clogged the openings that had been left in the cofferdams at construction sites. (The cofferdams were used to divert water from the river bed during construction.) Clogged openings caused the water to rise and flow over the tops of the cofferdams, thus flooding the construction areas. This, in turn, caused the loss of or damage to equipment. Refusing to regard the flood as a natural disaster beyond anyone's control, certain contractors sued the company, with litigation continuing for a number of years. Moreover, the power company had what it regarded as "bad experiences" with contractors on the big Bridgewater-Lake James project that was begun soon after the 1916 flood.[1]

W. S. Lee and his colleagues decided in 1924, at any rate, that the company could build its own generating plants and dams more cheaply and efficiently than any outside contractor. Before a description of the process in some detail, as it was carried out in the construction of the large-scale steam plant authorized by J. B. Duke not long before he died, perhaps a thirty-five year overview of the construction department's work, an overview written in 1960, might add some understanding of the significance of the "do-it-yourself" policy.

Duke Power's construction department was first headed by Arthur Carl Lee, younger brother of W. S. Lee. A. C. Lee, however, took a leave from the power company in 1927 to head the construction company that built Duke University's Tudor Gothic buildings on the West Campus and to do so employing a large number of techniques and practices perfected at Duke Power. Charles T. Wanzer, therefore, became the head of the department and remained in that position until 1958, when he became the company's chief engineer.

From 1924 to early 1960, the construction department built a total of seven steam-electric generating plants (with twenty-nine generating units) and four hydroelectric plants. (Important new projects underway in early 1960 will be discussed later.) Gradually, the construction department by itself became a large business. By 1960 it supervised about 1,000 workers and had equipment worth over $2 million. A then-recent issue of the *Engineering News Record* contained a list of the top 77 contractors in the United States for 1958. If Duke Power's construction department had been in the contracting business, it would have ranked 6th in the nation based on the dollar value of work in progress and contracts awarded.

C. E. Watkins, author of the 1960 overview and Wanzer's successor as head of the construction department, declared that Duke Power had been "singularly blessed, especially at the supervisory levels" by having men of excellent capability and "know-how." While many people thought of construction workers as itinerants, Duke Power had been able to keep the greater part of its labor force, especially the key people, continuously employed for many years.

In early 1960, for example, the construction department had 84 men with 10 to 14 years service, 46 men with 15 to 24 years, and 6 with 25 to 45 years. That made a total of 136 employees with more than 10 years of continuous service. Also there were more than 300 employees with continuous service of from 5 to 10 years.

What were some of the results from this situation? According to figures reported to the Federal Power Commission, the average cost per kilowatt of all new, non-Duke steam-electric generating plants put into service from 1948 through 1957 was $166. The average cost for new Duke plants in the same period was $103 per kilowatt—$63 less than the average non-Duke cost. If Duke Power had constructed plants at national average cost in that period, the installed capacity would have cost the company $19,530,000 more than was actually expended. (Watkins did not point out the fact, but Duke Power customers were the beneficiaries of those savings through lower rates.)

The cost of additions to existing plants from 1948 through 1957 told the same story. If Duke Power had been required to pay the national average cost per kilowatt, its plant additions would have cost an extra $39,840,000.

In a comparison of national cost figures for new steam plants for the year 1957, Duke Power's closest rival was the Gallatin Plant of the Tennessee Valley Authority. Its construction cost per kilowatt was $156, some $47 per kilowatt more than the cost of the two units in Duke Power's newest steam plant. Wrapping up his cost comparisons, Watkins noted that if one figured the costs of all steam plants, both new and additions, brought into service in the United States from 1926 to 1957, one found that if Duke Power had been required to pay the national average cost per kilowatt, then its steam plants would have cost $105,391,000 more than they actually did cost.

"Our construction cost record is the envy of the entire industry," Watkins declared. "A week seldom passes that someone from some other utility doesn't ask the question, 'How in the world do you do it?'" Watkins believed the answer to be simple: close cooperation between the construction, the design, the operating, the maintenance, and the procurement agencies as well as other departments within the framework of the company. Yet the principal answer, he suggested, was preliminary planning, able supervision within the organization of the construction department, and a large group of experienced, highly skilled, and loyal workers who had the best interest of Duke Power at heart.[2]

In 1925–26 no one could possibly know, of course, just how the company's new "do-it-yourself" construction policy would work out and what its long-range consequences would be. Still, the construction of the first large-scale steam plant by the company's own workers does take on more meaning in light of the 1960 retrospective overview.

W. S. Lee, Richard Pfaehler, David Nabow, and other engineers involved in planning and building the mammoth new steam plant were justly proud of their achievement: they placed a 100,000 horsepower plant in operation within nine months after ground was broken for the plant and within less than a year after the purchase of the site and the beginning of work on the design of the plant.

Although Pfaehler and others wrote reports on the project for specialized or professional journals, Lee gave a talk about it that was more suitable for a general audience. He began with an Uncle Remus story in which the rabbit "clumb a tree." When some listener protested that rabbits could not climb trees, Uncle Remus replied that "this rabbit was 'bliged to." Lee's point was that, in view of the drought of 1925 and resulting power shortage, Duke Power felt "'bliged" to respond as quickly and effectively as possible.

Lee explained that after J. B. Duke's death in October 1925, his associates in the company made their final decision about the size and location of the plant in the following month. They also decided to name it Buck Steam Station in tribute to the late president—a name, incidentally, that they would never have dared to use during his lifetime.

Since large quantities of water are essential even for steam plants, the plant would be built on the Yadkin River close to Salisbury, North Carolina. The construction department's first job was to build a railroad two miles long for the transport of building material and of coal after the plant was ready to be put into operation. At the same time, a camp for the construction workers was built and work begun on permanent quarters—forty-two houses—for the personnel of the plant once it began operations.

About 150 men began work at the site in January 1926, and the work force gradually grew to approximately 700, with double shifts on some portions of the work. The first concrete was poured in early March, after a most exacting job of excavation. This was so, Lee explained, because the plant was sited immediately on the river bank, and excavations had to be made to bed rock considerably below the level of the river's bottom.

With what Lee described as "splendid co-ordination" between the various departments of the work on the boiler room, turbine room, coal-crushing and pulverizing plant, and other parts, the work progressed so rapidly that on October 8, 1926, the first boiler was fired up, and four days later the first big turbine turned. Thus the plant began to deliver power in three days less than nine months after the groundbreaking.

Each of the two main turbo-generators weighed 450 tons, and the speed of the turbines was 1,800 revolutions per minute. The foundation under the turbo-generators consisted of structural steel encased in concrete, with the foundation for each being detached from the rest of the building so that no vibration could be transmitted from one machine to another and thereby disturb the smooth operation of the rotating element.

Current was generated at 13,200 volts and then stepped up to 101,200 volts in outdoor-type transformers; from them it was fed through the transmission system of the company; stepped down by transformers; and delivered to the various industries and towns on the company's lines in the Piedmont Carolinas.

Lee boasted that the most improved and economical power-plant technology of the day was utilized in the Buck Station. There were four huge boilers (34' wide, 26' long, and 58' high) that occupied space equivalent to a 6-story building 214' long and 135' high. Combustion chambers formed the lower part of the boilers and were 30' high. The pulverized coal was fed into the combustion chambers with carefully calculated quantities of air so that combustion was practically complete. Two smoke stacks towered 207' above the mean water level of the river.

When operated at full load, Buck plant consumed approximately two dozen 50-ton car loads of coal a day. Storage facilities covering more than six acres held 80,000 tons of coal. It was delivered from the cars to the crushing plant, then dried and pulverized on the way to the bunkers above the boilers.

Buck Station, like similar steam plants across the nation, used much water. For every ton of coal burned, 350 tons of condensing water were required. The object of the condenser was to create a vacuum into which the steam exhausted so as to draw out the full value of the steam and conserve the pure water for continued use through the boilers and turbo-generators. There were nearly forty miles of tubes in each condenser.

Lee concluded by noting that the building of the Buck Steam Station had relieved the company of operating older, smaller, and much less efficient steam plants to provide the generating capacity to carry the company's constantly increasing load while new hydroelectric plants were being built.[3]

Lee's closing reference to the building of new hydro plants was not mere rhetoric, for in early 1927 the company began construction of Oxford Hydro Station. Located on the Catawba some seven miles below Taylorsville, North Carolina, the 36,000-kilowatt plant began operations in the spring of 1928. Oxford's dam, like some others Duke built around the same time, had floodgates, for on the earlier dams no provision was made for the regulation of the reservoir pond level by means of flood gates on the crests of the spillways. Experience in dealing with the Catawba had suggested, however, that it would be advantageous to have such control works. More important, perhaps, was the fact that Oxford was Duke Power's last hydro plant to be built on the Catawba until 1960.[4] Not only were suitable hydro sites on the intensively-developed river becoming hard to find, but the new technology used in steam plants made them increasingly attractive from an economic point of view.

Just about the time the Buck Steam Station was being completed and the Oxford hydro plant being started, Thorndike Saville, a professor of hydraulic engineering at the University of North Carolina (who was also the chief hydraulic engineer for the North Carolina Department of Conservation and Development) took a revealing overview of "The Power Situation in the Southern Appalachian States." The article has interesting things to say about Duke Power (although the company was still referred to in the article by its old name of Southern Power) and sheds light on the electric utilities of the entire region.

The Southeastern "Power Province," according to Saville, had not been and probably would not be dependent to any extent upon power sources outside of its own borders. The province was described as an economic entity, the vast natural and human resources of which had begun to be developed on an extensive scale only during the previous decade. "This development is predicated chiefly upon three factors," Saville maintained. "Power, which is cheap, widely distributed and of considerable magnitude; labor, which is of Anglo-Saxon stock, intelligent and largely non-unionized [which simply ignored the large population of African Americans in the region]; and adequate transportation facilities by means of greatly improved railroads and highway systems."

Saville believed that the progress of power development had perhaps been the most tangible evidence of the area's growth. Moreover, it was that which had caught the attention of other parts of the country.

The power companies of the South had succeeded, he maintained, despite natural difficulties, floods, inadequate capital, pessimistic predictions of failure because of lack of a market for power, and a general public indifference. The early companies had been largely the handiwork of a few southern men with far-seeing vision; with their own capital and that of a few associates, they had persisted in the face of difficulties such as had been met and overcome in no other portion of the country, Saville argued.

After reviewing the early history and growth of the Southern Power Company, Saville noted that in many instances it had been a leader in the industry by adopting new methods of generation and transmission. "Always pioneering, always aggressively independent, under the exclusive control of a few men of means who never found it desirable to place any stock on the open market and who consistently reinvested their profits in extending the service of the company, more or less indifferent to attempts upon the part of public authority or private groups to regulate or coerce [it], the Southern Power Company," Saville declared, "has occupied a unique place in the power development of the South."

Saville went on to point out that the Duke/Southern Power Company, so highly individualistic and preeminently southern in control and personnel, stood out in the late 1920s as "an outstanding exception to the general tendency toward the consolidation of power properties under the supervision of the great public-utility management corporations with headquarters in New York and Chicago."

He pointed out that the genesis of the Georgia Power Company and various others that he discussed was different from that of Southern Power. He also noted that the 110,000-volt line connecting the Georgia company's Tallulah Falls plant with the lines of the Southern Power Company in 1913 was "the first step in the present inter-connected power province of the South."

The Carolina Power and Light Company, supplying electricity to the eastern portions of the Carolinas and to the Asheville, North Carolina, area, had become associated with a major holding company, Electric Bond and Share. Saville described C P & L as unique in-its manner of growth, for it had been built largely upon the consolidations of power properties begun by local capital and then taken over by C P & L after the local company's failure or after experience demonstrated an inability to give adequate service.

The Federal government's Wilson Dam at Muscle Shoals on the Tennessee River was destined later to play a key role in the New Deal's policy concerning electric power. Saville maintained that if it were not for the interconnected system of the southern power province, the Federal government would have been paying interest on the more than $36 million invested at Muscle Shoals without receiving any income. As matters stood, however, in 1926 alone the Alabama Power Company paid almost $860,000 for power that was produced at Muscle Shoals and then distributed across Southeastern states. Through the interconnecting systems, power produced at Muscle Shoals was, in effect, being used not only in Charlotte but in Raleigh, some 600 miles away.

The flexibility of the interconnected system made it possible, when there was a light load on the lines of one company's system, to send power to another company where a temporary heavy load existed. This lessened the need for expensive stand-by equipment to meet peak demands. Furthermore, rains on the territory of one system could produce surplus power that could then be transmitted to the territory of another system where no rains had fallen.

Saville noted that, with the exception of Southern Power, the larger power companies of the South had, within the previous few years, all become connected with some one of the larger management or holding companies. These companies exercised their control in a number of ways, such as by controlling the common or voting stock either by actual ownership of a majority of shares or of a large enough block of shares to have effective control.

The management or holding companies supplied consulting engineers, arranged for additional financing, supervised operation policies and frequently acted as a purchasing agent or as a subsidiary construction company to build plants and transmission lines for all of the associated companies. (What Saville did not explain was that all of the activities mentioned were highly lucrative for the management or holding companies and indirectly added to the cost of the electricity produced.)[5]

Duke Power then, as emphasized in the above analysis, hewed to its own course. With the industrial component of the Piedmont Carolinas' economy flourishing in the late 1920s—even as the agricultural component languished—expansion continued to be the order of the day for the power company. In early 1928, C. I. Burkholder, Duke Power's general manager and a vice president, reported on the continuing projected need for new generating capacity to the company's president in New York, George G. Allen. Burkholder first observed that power requirements for 1927 had been a shade over 18 percent greater than for 1926. Assuming that requirements in 1928 might be as much as 15 percent greater than in 1927, Burkholder also noted that the increase in central station energy over the whole nation in 1927 was only 8.3 percent while Duke Power's increase was over 18 percent. He reminded Allen that the company was still purchasing surplus power from other utilities but that it was "necessary for us to backstand purchased contracts by our own resources. . . ." In short, Burkholder concluded that it had become increasingly apparent to him that it was "necessary to proceed with the [creation of] additional steam generation capacity."[6]

Allen replied that expecting a continuous 15 percent annual increase might be a bit too optimistic. Nevertheless, he thought also that "we are all now fully alive to the necessity of the necessary preparations to enable us to meet any reasonable demands which can possibly be made upon us. . . ." In short, he thought the company should proceed promptly with the construction of a new steam plant.[7]

Accordingly, in 1928–29 Duke Power built the first two units of Riverbend Steam Station on the Catawba about twelve miles northwest of Charlotte. Using water from the reservoir of the Mountain Island plant, Riverbend was destined in future years to acquire numerous additional generating units and to become one of Duke Power's important mainstays for the generation of base load during many decades.

During the construction of Riverbend in 1928–29, W. S. Lee still stood before the engineering profession and the public at large as the chief spokesman for Duke Power. (George G. Allen, like his predecessor, J. B. Duke, studiously avoided public speeches and publicity in general.) Presenting a paper to the southern group of the American Institute of Electrical Engineers in the autumn of 1928, Lee used Duke Power's experience to emphasize the "advantages and economies which may be obtained by coordinating both hydro and steam power stations of various kinds, and electric transmission and distribution lines through a unified and centrally controlled system."

Sharing the widespread economic optimism of the time, Lee foresaw a promising future for the electric utility industry as well as American business in general. "The demand for power and light in the home, on the farm and in industries," he declared, "is increasing steadily and new fields are being opened constantly." Duke Power, accordingly, was pushing ahead with an expansion of and improvements in its system to meet future service requirements.[8]

While Lee was a prodigious publisher of articles in the leading engineering and trade journals, he had even earlier in his career earned a reputation in his field as an innovator, a creative technologist. Early in the century porcelain insulators had been mounted on large wooden or metal pins, but that method of mounting insulators was both costly and mechanically weak. Lee, while still a young engineer, invented a special insulator pin and base made of cast iron, which quickly replaced the earlier wooden device. Known as the "Lee pin and base," it was still in use in the 1990s.[9]

There were numerous other technological innovations for which Lee and other Duke Power engineers became known throughout the industry. That fact coupled with Lee's ability to communicate, to administer, and to lead should help explain why he was elected in 1930 as president of the American Institute of Electrical Engineers, one of the first Southerners to gain that recognition. As president, Lee managed in 1931 to bring the annual convention of the electrical engineers to the South; it was only the second time that had happened in the forty-seven-year history of the organization. By then, however, the economic scene was by no means as bright and promising as it had been in the late 1920s.

The famous Wall Street crash of late October 1929, did not cause the Great Depression of the 1930s. The crash did, however, signal the beginning of a prolonged, baffling series of sour and unhappy economic developments, not only in the United States but finally in Europe and elsewhere.

As the economy spiraled deeper and deeper into depression, an already-old fight about the Federal government's proper role in the production and distribution of electricity took on a new intensity. As mentioned earlier, the

Federal government built a dam, hydroelectric plant, and nitrate plant at Muscle Shoals on the Tennessee River during World War 1. After the war, the investor-owned utilities, their allies, and other groups that feared further government intrusion into the economy called for the sale or lease of the Federally owned facilities to private companies. Progressive reformers in Congress, however, such as Senators George Norris of Nebraska and Robert LaFollette of Wisconsin, strenuously demanded that the Federal government retain ownership with a view toward making "public power" an alternative to private or investor-owned power companies.[10]

When Senator Norris found a champion in President-elect Franklin D. Roosevelt early in 1933, the reform-minded advocates of public power won the day. In May 1933, Congress created the Tennessee Valley Authority as a multi-purpose public corporation—and it was soon on the way to doing on a massive interstate scale many of the things, and more, that the Southern/Duke Power Company had been doing on the Catawba—on a capitalistic basis—since 1904.

Needless to say, William States Lee, George G. Allen, and countless others associated with investor-owned utilities were adamantly opposed to the TVA. They saw it, understandably perhaps but incorrectly as matters finally turned out, as an entering wedge for government ownership of the means of production—a first step toward socialism. In a radio address in September 1931, even before the advent of the New Deal and TVA, Lee warned that there were widespread efforts underway to "make a political football" of the electric utility industry and to develop "a sentiment through which public ownership of electric utilities might be used as an entering wedge in that it may lead to the socialization of all industry. . . ." Lee thought it significant that the "agitation and the widespread and adroit propaganda in behalf of government ownership has had its source not among users of electricity and other utility services but among those who have had selfish interests, political or otherwise, to serve."[11]

Despite Lee's forebodings and the worsening economic scene, honors kept coming his way. In 1920 delegates from seventy-one engineering societies in the nation launched the American Engineering Council, an ambitious effort to unify the multifaceted profession. The Council's first president was none other than Herbert Hoover, already famous for his leadership of famine-relief efforts in Europe after World War I and soon to become even better known as Secretary of Commerce and then President.

The American Engineering Council elected W. S. Lee as its seventh president for 1932–33. In what seems to have been Lee's presidential address delivered late in 1933, he spoke on the broad topic of "Water Power Development." After tracing the beginnings of hydroelectricity in the 1880s and 1890s, Lee turned to advances within the industry in the more recent period.

Noting that generating units had increased steadily in size from 31,000-horsepower units in 1916 to the 115,000-horsepower units under construction in 1933, Lee pointed out that transmission voltages had similarly increased: from a maximum commercial voltage of 25,000 volts in 1896 to the 275,000-volt (275-kilovolt) lines under construction in 1933.

Paralleling the increases in the size of individual units and of plants, the efficiency of unit, plant, and system had been steadily improved. Lee noted that modern turbines used in conjunction with the latest type of waterway showed efficiencies at rated load of well over 90 percent even for low-head plants and relatively small units. For larger units, efficiencies of over 93 percent had been obtained.

Transmission efficiencies due to unified system operation and proper design had been vastly improved. Early in the century a plant efficiency of 70 percent had been considered excellent. By 1933, however, 89 or even 90 percent efficiency was attainable. Coordination of steam and water power resources, coordination and control of water storage facilities, coordination and cooperation to secure maximum economic use of interconnection facilities between systems—all those had resulted in vast improvement in system efficiency.

Lee then turned to a subject he knew best, the Duke Power Company, to illustrate power-system development. Starting with one plant producing 3,300 kilowatts in 1904, the company by 1933 had 504,000 kilowatts in hydroelectricity and 281,000 kilowatts in steam facilities. The annual power generation was in excess of 1.4 billion kilowatt-hours. Transmission networks covering the industrial area of the Piedmont Carolinas consisted of over 4,000 miles of circuits, approximately half of which were 100-kilovolt lines. Interconnection with adjoining power companies had proven both economical and mutually advantageous.

Lee singled out the hydroelectric development of the Catawba for special attention because twelve plants on the river utilized over 90 percent of the total head (fall) between elevation 1,200 feet and elevation 140 feet. In conjunction with water storage facilities, the twelve plants provided almost complete regulation and utilization of the river. The Bridgewater, Rhodiss, and Oxford developments on the upper reaches of the river and the Catawba developments located below the junction of the South Fork had large storage reservoirs for the retention of flood waters and excess river flow. The total reservoir capacity when utilized over the plants below amounted to 277 million kilowatt-hours.

The development and coordinated operation of this vast storage with the run-of-the-river plants, steam, and interchange facilities had been, Lee explained, the result of careful study and a definite plan of procedure. While the first development had claimed a plant efficiency of about 70 percent, many of the newer developments showed plant efficiencies at full load of 88 to 89 percent. Moreover, for 1933 efficiency of transmitted power from generator terminals to the customer's meters was over 86 percent.

Conceding that the Duke system was but one of many across the nation, Lee suggested (not altogether correctly) that the story of one system was the story of all—"maximum utilization of natural resources for the common benefit of power consumers and power producers."

Centralization and coordination of power resources had been achieved solely for economic reasons: to give uniform and reliable service; to maintain and even lower rates in the face of increasing costs; to make the investment sufficiently attractive that the properties could be financed; to afford the support of organizations highly trained in efficient planning, constructing, and operating the systems; and to engage aggressively in the development of additional power uses by domestic, commercial, and industrial consumers.

As of 1933, Lee noted, the aggregate capacity of hydroelectric equipment installed in the United States was more than 15 million horsepower, generating about 35 billion kilowatt hours annually. Also, the rate of increase for hydroelectricity was still more rapid than for power generation by steam.

After such a positive recital of the industry's progress, as well as of Duke Power's, Lee, with an eye on the Tennessee Valley Authority and other New Deal measures, ended on a sternly cautionary note:

> While our great power industry as now constituted has contributed so much to the needs of mankind and has been such a great factor in the upbuilding of our nation, sinister shadows tend to stunt, if not to kill, future normal growth. We are faced not with constructive regulation for the greatest good to both power producer and power consumer, but with destructive governmental regulation, destructive governmental competition, and destructive taxation. Unless saner judgment prevails, this great industry will see its handiwork sacrificed to political ambition, political oppression, and political aggression—to the lasting benefit of none and to the detriment of all.[12]

Lee did not know when he prepared the above piece that it would be his valedictory address. It was published posthumously, since on the morning of March 24, 1934, Lee suffered a cerebral hemorrhage (stroke) while at breakfast in his Myers Park home in Charlotte. Sixty-two years old, he died at his home that evening.

With the untimely death of W. S. Lee, Duke Power had lost the last of the three men whom the company hailed as its principal founders. Dr. W. Gill Wylie had died at age seventy-four in 1923, some two years before the death of J. B. Duke. While the company would later honor Lee's role by naming a major steam plant for him, perhaps his finest monument was also the Duke Power Company itself. J. B. Duke would have been the first to admit that Lee had played a major and indispensable role in the building of one of the nation's trail-blazing and pace-setting power companies.

Just about the time that W. S. Lee died, his warnings about New Deal threats to the viability and profitability of investor-owned electric utilities ceased to be a theoretical matter. Duke Power and the Duke Endowment had a head-on collision with the New Deal beginning in mid-1934. Greenwood County, South Carolina, applied to the New Deal's Public Works Administration for a loan of $2.6 million in order to build an electric-power plant on the Saluda River. Since Greenwood County was in the heart of the territory served by Duke Power, George G. Allen, W. R. Perkins, Norman Cocke, and others involved with both the power company and the Endowment were understandably irate when they learned from the newspapers that the Board of Public Works, chaired by Secretary of the Interior Harold Ickes, had voted four to two in favor of making the loan.

George Allen immediately wrote to Ickes protesting the loan and requesting that Duke Power be given at least a hearing. At the same time W. R. Perkins informed the Endowment trustees that the loan created a precedent for the granting of similar loans by the Public Works Administration to every county or municipality that might apply; that Duke Power, which had to pay taxes, could not compete with publicly-owned power plants that could borrow government money at low rates of interest and paid no taxes; and that the consequent reduction of income for Duke Power would reduce the amount available to the beneficiaries of the Endowment.

In George Allen's letter to the beneficiaries of the Endowment and a similar letter to the *New York Times*, which gave considerable coverage to the matter, he began by quoting from that portion of the indenture where J. B. Duke explained his plan to have the revenues from the water-power development (Duke Power) used for the "social welfare" of the people in the territory served just as the developments themselves promoted their "economic welfare." Allen then noted that since the Endowment's establishment, it had paid over $19 million to the beneficiaries, nearly 58 percent of which had come from Duke Power securities. "The Federal Government," he warned, "is pursuing policies, which, unless abandoned, will seriously cripple, if not destroy, the Duke Power Company."

Allen next explained that, after protests from him and others, Secretary Ickes had finally consented to have his committee on electric power hear the representative of Duke Power. "Unfortunately, however, we cannot feel sure that the merits of the matter will prevail," Allen continued. "We say this because of the well-known views of the element in the present Administration which is positive and outspoken in the belief that the government should own and operate the utilities."

Echoing earlier arguments by Perkins, Allen insisted that no investor-owned utility could successfully compete with a government-backed utility that had ample, cheap capital at its disposal and which operated without the burden of taxes and code restrictions (referring to the production codes of the National Recovery Act). "Therefore, government can easily use its revenues to destroy private utilities," Allen argued, "if it chooses to do so." He wondered why the government should use taxpayers' money to duplicate and thus ruin private business, in whose securities immense private funds had been invested. "Such a course is destruction," he insisted, "which makes for depression, not recovery."

Allen pointed out that electric rates were regulated by commissions chosen by the people or their representatives in the various states. Why, then, was there a need for government ownership in the utility field? Ownership of the utilities by the government, he warned, would only augment bureaucracy and increase the public debt as well as taxation. Then in conservative words that would probably have found a larger and more appreciative audience in the 1990s than they did in 1934, Allen concluded by saying that government ownership "kills private initiative, which alone accomplishes true permanent progress and prosperity." Did not history reveal that such had "been the record of all nations which have thus enhanced government by the suppression of the individual?"

In Allen's letter to the *New York Times*, he also pointed out that some 30 percent of the federal funds for the Greenwood County project would be given outright while the balance would be lent at long term at 4 percent interest. While the Greenwood project would pay no taxes, out of every dollar collected by Duke Power in South Carolina, twenty-seven cents were paid out in taxes. Moreover, Duke Power, in response to the depression, had voluntarily abandoned contract minimums for large industrial users of its power and made rate reduction at an annual aggregate cost of $2.5 million. "But existence, not rates," Allen concluded, "is at stake."[13]

As Allen had feared, Secretary Ickes' board of review approved the grant-loan but also required Greenwood County to gain clearance from the South Carolina courts to issue revenue bonds for the purpose of building a power plant. When the state's top court, rejecting the arguments of the Duke lawyers, gave the go-ahead signal for the project, Duke Power's lawyers asked for and obtained from the federal courts an injunction against further action in the matter by Greenwood County and the Public Works Administration.

When the matter reached the United States Supreme Court late in 1936, it ordered that the case be retried in the lower courts. Thus the whole affair dragged on until December 1937, when the case was again before the Supreme Court. "Bare of legal technicalities," as the *New York Times* reported, the point of the case was "whether the [federal] government was legally entitled to finance municipal power plants in competition with private business." Restating arguments that Allen and Perkins had made earlier, the counsel for Duke Power, W. S. O'B. Robinson, was described as "waving his arms and shouting," when Chief Justice Charles E. Hughes leaned forward and said with a smile, "Will you restrain your voice? It will help your argument." Although Robinson reportedly spoke less vehemently thereafter, that did not help his case, for the Supreme Court ruled unanimously against Duke Power and in favor of the government's program to lend and grant money to publicly owned electric plants even in competition with private enterprises. Secretary Ickes happily declared that the Supreme Court's action had ended a three-year fight and released almost $110 million to sixty-one projects being held up by injunctions in twenty-three states.[14] (Ironically, in 1966 the citizens of Greenwood County voted by a substantial majority to sell the county-owned electric system to Duke Power.)

Aside from the disappointment for Duke Power in losing what it regarded as an important case, the whole matter of keeping the company afloat in the stormy economic seas of the Great Depression required adroit maneuvering. Fortunately for Duke Power, Norman Cocke, E. C. Marshall, Charles I. Burkholder (who succeeded W. S. Lee as chief engineer), George Allen, and others proved to be an able crew. Industrial use of electricity plummeted as the depression deepened in the early 1930s. As early as May 1930, Burkholder notified all of the company's textile-mill customers that Duke Power had decided to remit all minimum charges for power that might accrue because of mill curtailment, that is, halted or slackening operations. "This action is taken on our part," Burkholder explained, "because we believe it will, to some extent, aid the mills in restoring the industry to a sound, economic basis, thereby resulting in mutual benefit not only to the mills and the Power company but to business generally in this territory."[15]

Burkholder was not the only Duke Power official who strove to do everything possible to mitigate the economic and social consequences of the textile industry's crippled condition. Norman Cocke, company attorney and vice president, worked closely with mill owners in the Piedmont to help them keep factories open and unemployment down. According to the later account of a Duke Power officer, "Industries were kept in business that would have gone bankrupt and many people were kept at work that would have suffered, and friendships were made which help explain why Mr. Cocke is probably the best known and best liked man that has ever worked with this Company." When all the banks were closed in Gastonia, North Carolina, for example, Cocke helped arrange for Duke Power to lend an industrialist there $25,000 to open a bank.[16]

With construction of new plants out of the question (and remaining so until 1939), Duke Power focused its attention on residential, municipal, and small business customers—groups that had been of only secondary concern in the early days. There was a concentrated effort to sell electric appliances for the home, and the public responded by replacing ice boxes with refrigerators and also purchasing toasters, percolators, fans, and other appliances. After a rate reduction, the company reported in 1932 that there had been "an increase in the volume of sales of household appliances." Electricity was no longer the expensive luxury that it had been in the 1890s, and one housewife declared that, "It's such a relief to not feel that you should turn off the lights behind you every time you leave a room, and to be able to plug in any appliance that you want to use anywhere in the house...."[17]

In 1935 the company made the already low rates even lower with a special rate for electric water heaters. As the capstone for the drive to sell the new water heaters, Duke Power now charged only 1½ cents per kilowatt hour for the first 200 kilowatt hours of electricity used each month and one cent per kilowatt hour for all electricity used in excess of 200 hours.

Aside from advertising electrical appliances, the company adopted a merchandising policy that included small cash downpayments with credit terms extended over a few months in the case of small appliances up to two years for the more expensive ones. The same type of policy was used in house-wiring campaigns: the customer selected the wiring contractor who made an estimate of the cost; then the company paid the contractor, with the customer reimbursing the company in twenty-four monthly payments without interest.[18]

In addition to the intensive effort to build residential and commercial load, Duke Power cultivated farmers in the 1930s as never before. Because of the expense of building transmission lines to widely scattered and isolated farmhouses, electrical service to farmers in the United States lagged behind that which was available for them in Germany, France, Holland, and the Scandinavian countries.[19] President Franklin D. Roosevelt moved to begin changing matters by issuing in 1935 a presidential order creating the Rural Electrification Authority as a relief agency. Then in the next year Congress enacted legislation giving it a permanent statutory basis. By 1940 REA had granted loans of over $321 million for rural electrification co-operatives.[20]

Actually, Duke Power had begun to serve many Carolina farmers even before the creation of the Rural Electrification Authority. The New Deal's rural activism, however, inspired Duke Power to move more aggressively. In early 1937, E. C. Marshall, who had long headed the Southern Public Utilities subsidiary before its merger with Duke Power in 1935, came up with the idea of creating an agricultural engineering department within the company. To head the new department, Marshall recruited Marvin T. Geddings, a South Carolinian who had graduated from Clemson University in 1930.

When Geddings asked Marshall what an agricultural engineer was supposed to do, the latter replied, "I don't know a damned thing about it but it's up to you to find out." Thereupon, Marshall assigned Geddings to the company's office in Spartanburg, South Carolina, with the responsibility of working the entire Duke Power service area from the Virginia line to the Georgia line.

At that time (1937), the company had approximately 150,000 residential customers, of whom 50,000 were rural. (A "rural" customer was not necessarily a farmer, of course, given the dispersed pattern of Carolina textile manufacturing.) Geddings also later recalled that approximately a fourth of all residential customers paid the minimum charge of 80 cents per month.

Marshall may not have known exactly how the company's new "agricultural engineer" was to proceed, but he did have certain large objectives in mind. First, he hoped to build good will for the company by having some one to work directly with rural people in the service area. Second, he hoped to add load to the system through the application of electric service in various farming operations. And finally, Marshall wanted Geddings to work with and assist other agricultural agencies in their programs of rural community development.

Geddings spent his first year on the job getting acquainted with and discussing programs with agricultural leaders in the Carolinas, attending farm meetings, and visiting farm families to learn their practices and needs. In order to demonstrate how electrical appliances could be used by farmers, he arranged for the company to buy a second-hand enclosed trailer pulled by a 1932 Chevrolet coupe. Geddings loaded the trailer with a refrigerator, water heater, electric stove, small appliances, homemade electric chicken brooders and water warmers, dairy milker, milk cooler, bottle sterilizer, soil sterilizer, soil heating cable, and various other items.

Taking his trailer to meetings at schools, churches, fairs, and wherever farmers congregated, Geddings had a Duke lineman set a large transformer for the trailer to be connected to and a Duke Power home service advisor (usually a woman who had studied home economics) to help in demonstrating the equipment.

The work went well, for in the second year another agricultural engineer was added to the department. With his office in Greensboro, North Carolina, the territory was then divided, Geddings being responsible for the South Carolina service area and seven counties in western North Carolina and the new man covering the rest of the North Carolina territory, After World War II and continuing into the 1950s and later, the department continued to grow.

In pursuit of the objectives that Marshall had specified, Geddings and his associates worked closely with the agricultural extension services at Clemson and North Carolina State as well as with the departments of vocational agriculture at Raleigh and Columbia. Friendly working relationships developed quickly. In addition to meeting with adult groups such as those in the Grange and the Farm Bureau, the agricultural engineers spent much time with 4-H clubs and sponsored awards for 4-H electric projects. Joining forces with Carolina Power and Light and the South Carolina Electric and Gas Company, Duke Power helped erect "electric buildings" at 4-H camps. The Future Farmers of America also received much attention.

As for adding load to the Duke Power system, one of the early successes came in poultry farming. According to Geddings, by the end of World War II, practically all hand labor had been eliminated from poultry farming. Electrically-operated devices for feeding, watering, and litter removal allowed one person to handle 5,000 layers or 40,000 broilers. Moreover, use of electric brooders on a poultry farm was desirable from the load-factor standpoint because it was an around-the-clock load with the highest kilowatt-hour consumption during the winter months. Brooders also used off-peak power during the night.

Mechanization of dairy farms posed a larger challenge to the agricultural engineers, according to Geddings. The dairyman had to handle an average of four tons of material for every ton of milk produced; or, put another way, he handled twenty tons of material per cow per year. Mechanization and electrification, however, made it possible for the dairyman to increase the size of his herd from twenty cows per person in 1933 to a hundred cows per person by the late 1950s. (What Geddings was describing, of course, was only a segment of the revolutionary transformation of American agriculture that allowed fewer and fewer workers to produce more and more commodities. In other words, even as the number of farmers and farm workers shrank, agricultural productivity, aided to a significant degree by electrification, soared.)

In the late 1950s the agricultural engineering department began to push the use of mercury-vapor "dusk-to-dawn" lights in rural areas around dairies, poultry houses, homes, churches, and schools. The cost was $3.00 per month, and the off-peak usage was another plus for the company. Duke Power was the first utility in the Southeast to use the lights, and after a year or so of their promotion, an airline pilot told a company official that when he got to the North Carolina line, in flying from Virginia at night, he could cut off his instruments and follow the mercury-vapor lights to the Georgia line.

Geddings obviously developed many valuable contacts, for he proved able to help Duke Power's legal department with certain problems that arose from time to time in the state legislatures. Working through his long-time friends in the Farm Bureau and the State Grange, Geddings, in short, had acquired a certain "clout." In 1955 the company asked him to work as a legislative agent, that is a lobbyist, in the South Carolina legislature, along with his regular duties. Then in 1972, three years before his retirement, Geddings was named as an assistant vice-president to work only with the South Carolina legislature. Obviously, Marshall's plan for an agricultural engineering department had worked out in all sorts of unforeseen ways.[21]

Agricultural engineering was not the only important innovation for Duke Power in the late 1930s. Just as the company's employment of Dr. Frank M. Boldridge back in 1923 had marked the beginning of an environmental and water quality program, so did the establishment of Duke Power's forestry department in 1939 signal an important commitment to another aspect of the environment: reforestation, water conservation, and the prevention of soil erosion.

As has been mentioned earlier, from 1904 onward the company had attempted to increase its revenue by farming certain tracts of land that had been acquired in the process of building reservoirs and dams. In general, farming proved no more profitable for the power company than it did for millions of Southerners, white and black, especially after about 1921. By the 1930s, of course, southern agriculture in general, and cotton growing in particular, had become virtually hopeless.

The company's farm agents, who contracted with the tenant farmers who did the actual farming, reported to Norman Cocke in the legal department. Worried about the soil erosion that harmed the reservoirs and weary of the unprofitability of the farming operation, Cocke determined to try a new course. There was also a matter of company-owned forest lands that were unproductive. Thousands of trees were lost annually because of old age, insects (such as pine borers), disease, and forest fires.

After seeking advice from the School of Forestry at Duke University, Cocke hired Carl J. Blades, a graduate of the forestry school, as the company's first chief forester in 1939. Beginning with the company's land around Great Falls, Blades launched a reforestation and watershed management program that grew more elaborate and extensive over the years. Like so many other Duke Power managers, he would spend his full career with the company, retiring after thirty-eight years in 1977.[22]

Though Duke Power strengthened itself even during the depression through such innovations as agricultural engineering and the forestry department, the fact remained that the overall economy of the Piedmont Carolinas, as in most of the nation, spluttered and faltered throughout the 1930s. An economic upturn in 1937 was followed by a sharp decline in 1938. Roosevelt and the New Deal, in short, proved more successful in introducing certain important reforms than in helping to bring about economic recovery.

The increase in residential and commercial sales, however, helped Duke Power to survive the long depression. In late 1936 the company decided it should add another generating unit to the two existing ones in the Riverbend steam plant. Then finally, after a long hiatus, early in 1939 Duke Power decided the time had come for the construction of a new plant.

Cliffside Steam Station, located on the Broad River about eleven miles southwest of Shelby, North Carolina, went into operation in mid-1940, about a year and a half after construction had begun. The two initial units of the plant were intended to be used as a supplement for hydro power; but in actuality, of course, it was the steam plants, like Buck and Cliffside, that were destined to provide the system's base load, while increasingly the hydro-electric plants would be used for peak load.[23]

The construction of Cliffside proved to be indeed timely, for World War II began in Europe in September 1939. (The war's Asian phase had begun earlier with Japan's assaults on China.) While the American people were initially united in hoping to avoid involvement in the war, they gradually grew angrily divided into opposing camps: the isolationists versus the internationalists (who were dubbed "interventionists" by their opponents). Long before the bitter debate ended with the Japanese attack on Pearl Harbor in December 1941, however, an overwhelming consensus in favor of the military preparedness of the nation had been achieved. Thus, by 1940, World War II and American rearmament began to revive the nation's economy in a manner that the New Deal had proved unable to accomplish. In the Piedmont Carolinas, Duke Power would need all of the electricity that its plants could generate. In fact, the company had to add two new generating units to the Buck Steam Station in 1940–1942. Then the Federal government's wartime priorities forced a halt to utility construction for the duration of the war. Not until 1946 could Duke Power embark on a huge program of much-needed expansion.

This is hardly the place to attempt to summarize the economic impact of World War II on the South or even the Piedmont Carolinas. "By every index of measurement," one recent historian of the South has declared, "the South's economy took off during World War II." [24] The value added from industry tripled, personal income more than doubled, and, the same historian points out, the number of industrial workers grew by better than 50 percent. The vast infusion of capital during the war (much of it from the Federal government), and the increase of managerial talent and industrial skills "primed the South for sustained industrial growth."[25] Since the Piedmont Carolinas had much more of a head start toward industrialization than did so much of the South, the war simply accelerated changes that had long been underway.

For the Duke Power Company itself the war brought the same mixture of patriotic exhilaration and aggravating shortages that it did to most American businesses and homes. More than 1,000 Duke Power employees went into military uniform, and the company was forced to relocate and retrain personnel in order to maintain reliable service, which was more crucial than ever during the war.

In the nervous weeks and months after Pearl Harbor, cities across the nation practiced blackouts and emergency drills in case of enemy attacks. (No one then knew, of course, that the Axis powers—Germany, Italy, and Japan—would not be able to attack the North American mainland directly.) In Charlotte, for example, at 8:00 p.m. on a winter day in 1942, just a few months after Pearl Harbor, a defense sub-coordinator for public utilities gave a telephone message to certain Duke Power employees: "All hellz-a-poppin'."

It was a code message signifying that Charlotte was being "bombed" by enemy planes. Within minutes, Duke employees proceeded to mobilize according to the pre-arranged plan. Three substations called in to report that they were prepared, and all street-lightning switches were pulled. Six minutes after the alarm had sounded, an assistant line superintendent dispatched crews to major Charlotte hospitals, the Quartermaster Depot, a radio station, and one or two other places that had been damaged by "bombs." Soon Duke Power trucks were at the various locations with men making imaginary repairs, checking lines, and testing equipment."[26]

Most of the company's role in the war effort was not so dramatic, of course. Given war-time shortages and military priorities, procurement of essential supplies and parts became a special challenge. Therefore, the company's oldest subsidiary, Mill Power Supply, had a larger and more urgent role to play than ever. B. B. Parker had begun work for Mill Power Supply in 1936, so he had some experience before facing the problems brought by World War II. His adroit handling of those problems, however, and the increased importance of procurement after the war led to Parker's eventually becoming president of Mill Power Supply in 1960 and finally of Duke Power itself in 1976.

With Duke employees, like so many other Americans, planting Victory Gardens and buying war bonds and saving stamps, World War II produced a far different spirit and morale from what would happen in connection with the Vietnamese War some twenty-odd years later. The war in Europe ended in May, 1945, and, after the United States dropped atomic bombs on Hiroshima and Nagasaki, Japan, World War II ended in the Pacific in August, 1945. The glory years for Duke Power—the late 1940s, 1950s, and 1960s—were about to begin.

CHAPTER IV

THE GLORY YEARS: FROM THE LATE 1940s TO THE LATE 1960s

Sandwiched between the grim, semi-stagnant years of the Great Depression of the 1930s and a different but highly troubled period in the 1970s, the two decades from the late 1940s to the late 1960s could well be called the "glory years" for investor-owned electric utilities in general and for Duke Power in particular. This was not simply because the problems that began to assail the industry so relentlessly after the late 1960s made the preceding two decades look benign and good. There were concrete explanations for the reality of what came to be regarded as the "good old days" for power producers.

In the first place, energy consumption in general grew with the economy in the two decades after World War II, but electricity sales rose at an even faster pace. Compared with the price of other fuels, there was a dramatic, continuing drop in the real price of electricity—despite an overall tendency in the economy toward rising prices, that is, inflation.[1]

Much of the success in reducing costs derived from steady improvements in the generation of electricity. In order to realize greater economies of scale, the electric utilities built larger and larger generating stations. In the late 1940s, most coal-burning steam plants were under 100,000 kilowatts in size, but in the 1950s plants in the 500,000–1,000,000 kilowatt range began to be commonplace. And by the end of the decade there were a number of stations of over 1,000,000 kilowatts. In other words, the average size of a power station rose five fold between 1945 and 1965.[2]

Sales of electricity, measured in kilowatt-hours, showed strength even in recessionary years (such as 1957–58), and the industry could build large new plants with confidence in the marketability of what they produced. Providing increasingly cheap electricity, the industry faced costs that were either under control or actually declining, although prices within the economy as a whole were rising. A few, quite modest rate increases in the 1950s were followed by significant rate reductions in the 1960s. All in all, those years were indeed good ones for the producers of electricity.

Duke Power and the Piedmont Carolinas fully participated in the economic joyride, for the growth in the region's economy significantly outpaced the national average. The most pressing task facing the company after the war, however, was the unglamorous but essential matter of catching up on maintenance and repair. During the war years, private industry had a low priority for many materials and supplies, and as a consequence, as one historian has noted, "Duke Power's facilities had been spliced and patched and pressed back into service time and time again."[3]

Pushing a crash maintenance program to replace much that was replaceable, the company still faced a huge problem: war-time shortages did not end when the war did. This was where the procurement skills and experience of B. B. Parker and the Mill-Power Supply subsidiary played crucial roles. Seeking to make the required repairs with minimum inconvenience to customers, maintenance crews went out each Sunday morning, a time when the demand for power was low. Customers received advanced notice that the power would be off during certain hours, and generally service was restored before the time for church services.[4]

Since no generating capacity had been added to the Duke system since 1942, that was another matter of some urgency. Only a few months after the war's end, the company announced that it would add two new generating units—at a cost of $4,500,000—to the original two units of the Cliffside Steam Station, which had been put into operation in 1940. The addition would give the station a total of over 150,000 kilowatts.

Duke Power then began to announce and build new steam plants with remarkable rapidity. The first two units of the Dan River Steam Station—located on the Dan River about four miles southeast of Eden, North Carolina, and in the northern segment of the Duke service area—went into operation late in 1949 and in March 1950, respectively.

Then, at the southern end of the service area, the company built on the Saluda River about seventeen miles south of Greenville, South Carolina, a new type of steam plant that was described as the most modern and efficient in the South. Named for William S. Lee, the Lee Steam Station, which began operating in 1951, with a rated capacity of 90,000 kilowatts featured several firsts on the system: 1) the two units utilized a re-heat cycle in which steam was re-routed through the boiler for added energy and efficiency to drive the turbines; 2) the burners were designed to operate on either coal or natural gas; and 3) cooling towers provided supplemental cooling to condense the steam.[5] Altogether in the five years after World War II ended,

Duke Power spent about $200 million for new plants and the necessary related equipment.

While all the expansion went on, there were also some important shifts in the top management of the company. In 1949 George G. Allen, president since J. B. Duke's death in 1925, gave up the job to become Duke Power's first chairman of the board of directors. At the same time, Edward C. Marshall was named as the president of the company.

An amusing story about Marshall suggests that Duke Power, despite its long record of trail-blazing and innovation in the industry, was not always in the forefront. On several occasions prior to 1948, some of the personnel in the operating department, and especially W. J. (Buck) Wortman tried unsuccessfully to convince management of the advisability of installing two-way radio equipment on the Duke system. Marshall always balked, arguing, "It costs too much, and it won't save any money and if it does[,] just put it in my hand."

Since some neighboring utilities were already using mobile radio to great advantage, Marshall, after much discussion and some pressure from other company officers, agreed to try out the new system. Some time later when Marshall came in about another matter, Wortman seized the opportunity to promote more use of two-way radio. Marshall gave his standard reply, "If it saves the company money[,] put it in my hand."

Wortman then told Marshall about something that had happened the previous month. One of the Charlotte branch service trucks, recently equipped with radio, had slipped off a mud bank during a rainstorm and turned over. Trapped inside, the service man called on his radio to summon help. An ambulance came to take him to the hospital where he had recovered, and he was now back at work.

"Mr. Marshall," Wortman said, "radio probably saved that employee's life. Was that man's life worth any money to the company?" Marshall, who usually kept an unlit cigar in his mouth, replied, "Of course it was." Wortman fired back, "Put it in my hand." Whereupon Marshall promptly bit his cigar into two pieces, stormed out of the office, and was never again heard to make any further comments about Duke Power's use of two-way radio.[6]

When Marshall died in 1953, Norman Cocke became president, and W. S. O'B. Robinson was named as vice-chairman of the board of directors and chairman of the executive committee. (George G. Allen, seventy-nine years old in 1953, was obviously playing a diminishing role in the affairs of the company.) All of these top managers, it should be noted, had been closely associated with J. B. Duke in the early days of the power company, so there was still a marked continuity in leadership. Moreover, while George

G. Allen and W. R. Perkins (who died in 1945) had played prominent and powerful roles in Duke Power, the Duke Endowment, and Duke University, Norman Cocke went them one better: he was an original trustee of the Endowment, and at the same time that he was Duke Power's president in the 1950s, he served as chairman of Duke University's board of trustees. It was perhaps the high-water mark of the inter-locking relationships among the three major institutional legacies of J. B. Duke.

The days of the old-timers in the company were, however, numbered. In 1957 George Allen gave up the chairmanship of the board to become merely honorary chairman, and W. S. O'B. Robinson took on the job as chairman of the board and general counsel. Then in 1958 when Norman Cocke resigned as Duke Power's president, something new and significant happened.

The naming of William B. McGuire as Duke Power's president late in 1958 marked the transition to a new generation of company leaders—men who had not been closely associated with J. B. Duke or, for that matter, who had neither seen nor known him. Forty-eight years old upon becoming president, McGuire was a native of Franklin, North Carolina, in the mountainous, far western corner of the state. After graduating from Davidson College in 1930, he obtained his law degree from Duke University in 1933 and went directly to work for Duke Power. Conspicuously able and dependable, McGuire was carefully groomed for top management, as was shown by the fact that in 1957 he was named as special assistant to the company president (Cocke).

In addition to marking a generational transition, McGuire's accession to the presidency illustrates another important feature of Duke Power's history: the company consistently and carefully recruited its top management from inside its own ranks. J. B. Duke, in his indenture creating the Duke Endowment, had requested the trustees "to see to it that at all times" the power company "be managed and operated by the men best qualified for such service." Just how much actual influence this request has had over the years, no one can say, of course. The historical record up to almost the end of the twentieth century suggests, however, that those who have controlled the company and selected its leaders have, either consciously or serendipitously, acted in accordance with J. B. Duke's words.

In the late twentieth century, many of the nation's largest business organizations seemed to change chief executive officers, usually brought in from outside, at the drop of a hat. Certain highly mobile CEO's became famous as axe-wielders who could strengthen "the bottom line," whether by ruthless down-sizing or some other technique. Against that background, it comes as something of shock to note that Peter Drucker, one of the nation's top gurus in the field of business management, once declared, "It is an admission of bankruptcy... [for a company] to have to go on the outside to recruit top management."[7] What Drucker meant, of course, was that it was precisely the ability of an organization to recruit able people, to help develop their talents, and to prepare them for the responsibilities of leadership that was one of the hallmarks of successful business. By that definition, J. B. Duke had first succeeded in the power company, and then those he picked somehow passed along the secret—or knack.

Perhaps it was appropriate for the relatively young McGuire to lead Duke Power into the nuclear era. The civilian nuclear power industry began on Labor Day, 1954, when President Dwight D. Eisenhower gave the signal to begin construction on the nation's first commercial nuclear power plant. It was a joint venture near Pittsburgh, Pennsylvania, between a local investor-owned utility company and the recently formed Atomic Energy Commission. With the Eisenhower administration enthusiastically pushing its "Atoms for Peace" program, the Atomic Energy Commission played a dual role of both promoting and regulating the nuclear power industry.[8]

Although a nuclear power plant is an inordinately complex affair, the basic idea involved is relatively simple. A steel-reinforced concrete containment building houses the reactor, which contains a core composed of bundles of pencil-thin rods of uranium fuel. Nuclear reaction within the core heats water, which circulates through the plant's radioactive primary loop and transfers its heat to the non-radioactive secondary loop. Then the steam produced from the heated water in the secondary loop drives a turbine, which in turn is harnessed to a generator that produces electricity.[9] Technologically, a nuclear power plant is a world away from the first Catawba hydro plant of 1904, but, in the final analysis, both systems worked to produce electricity by making a magnetic field spin within coils of copper wire.

While the Atomic Energy Commission issued construction permits for commercial nuclear reactors to two utilities as early as 1957, Duke Power preferred a more cautious and conservative approach. In October 1956, it joined with three of its neighboring utilities (Carolina Power and Light Company, Virginia Electric and Power Company, and South Carolina Electric and Gas Company) to form Carolinas-Virginia Nuclear Power Associates, Incorporated. With the approval and assistance of the Atomic Energy Commission, this was a research and development organization which proposed to build a small experimental nuclear reactor at Parr, South Carolina, on the Broad River north of Columbia.

Construction having begun in October 1960, Parr Nuclear Station generated its first electricity in December 1963, and the 17,000-kilowatt plant began commercial operation in May 1964. Duke Power owned a 34 percent interest in the facility, which cost altogether about $46 million.[10] Sending teams of Duke Power engineers and other technicians in relays to the Parr plant, the company used it as a training school, one whose benefits would be tangibly demonstrated in the years just ahead.

Among the young Duke engineers who cut their nuclear teeth at Parr was William States Lee III. A grandson of the W. S. Lee who was J. B. Duke's chief collaborator, Bill Lee (who disdained the "III") was a native of Charlotte who graduated first from Woodberry Forest preparatory school and then, in engineering, from Princeton University. After four years in the United States Navy's civil engineering corps, the twenty-five year-old Lee went to work for Duke Power in 1955.

He was attracted to Duke Power not by the family connection but by the opportunity of working with and learning from David Nabow. When that brilliant and nationally recognized engineer, who had joined the company back in 1916, died in 1958, young Lee, never known for patience, promptly handed in his resignation. Lee later recalled: "But the guy who took my resignation said, 'Be patient, boy. We haven't had the funeral yet. Just calm down.' So I don't know what happened, but gradually I got really turned on by what was going on at the power company and found it very exciting and challenging."[11]

"What was going on at the power company," of course, was its careful, studied approach to nuclear power, and Bill Lee was destined to become one of the nation's most widely respected leaders in the nuclear power field. Named assistant to the chief engineer, C. T. Wanzer, in 1959, Lee became engineering manager upon Wanzer's retirement in 1962 and in 1965 was named as vice president for engineering and construction. Winning countless national awards along the way, he was destined to become Duke Power's chairman of the board and chief executive officer in 1982.

While nuclear power seemed clearly in the picture for Bill Lee and Duke after 1956, it was still quite a few years away—and the Piedmont Carolina's economy and population were growing at a rate that would not wait for nuclear power to become a reality. Duke's response to the soaring demand for electricity took two forms. One response was to add generating units at a steady clip: two more units at Riverbend Steam Station in 1952, two more at Buck in 1953, two more at Riverbend in 1954; and a third unit at Dan River in 1955.

The company's other response to the constantly escalating demand for electricity was to build a vast new steam plant on the reservoir that was soon to be renamed Lake Wylie in Gaston County, North Carolina. Plant Allen, named in honor of George G. Allen, went into operation in 1957. With two 165,000-kilowatt generating units, it was the largest investor-owned power plant in the Southeast at that time and cost more than $41 million. Including the Allen plant's 330,000 kilowatts, the total generating capacity of the Duke system had reached approximately 2.25 million megawatts.[12]

As W. B. McGuire became Duke Power's president in January 1959, therefore, robust expansion was clearly the order of the day. Under McGuire's leadership, the company proceeded to strengthen its mode of operation in a number of ways. With its need for capital far outstripping the resources that could be provided by the Duke Endowment, Duke Power began to present itself in a new way, both to shareholders and the public at large. Having long provided only a bare-bones, brief annual statement on the company's operation and financial condition, Duke Power in 1958 began to publish a handsome, glossy *Annual Report* that contained numerous graphs and pictures as well as interesting, readable accounts of important developments in the life of the company.

For example, the *Annual Report* for 1958, the first of the new comprehensive reports, noted that revenues from residential sales during the year had increased 9.5 percent over those for 1957 and accounted for nearly 40 percent of the total electric revenue. Industrial textile power revenues, on the other hand, had decreased 0.1 percent from 1957 and were a bit over 25 percent of the total. (While the report did not mention the fact, the recession of 1957–58 no doubt explained the slight drop in revenue from textile sales.) Other, non-textile industrial and commercial revenues both increased.[13]

The rated capacity of the company's electric generating plants in 1958 was 2,783,000 kilowatts. The maximum net demands (peak loads) on the system were 2,287,235 kilowatts in August (as air conditioning began to be increasingly adopted in the region) and 2,306,530 kilowatts in December.[14]

Duke Power's service area covered about 80 percent of the Piedmont Carolinas, an area of about 20,000 square miles lying between the Appalachian Mountains and the coastal plain. The region contained 55 percent of the metropolitan population of the Carolinas and about 40 percent of the non-metropolitan population, in a land area that comprises about 30 percent of that of the two states.

Stressing a theme that would continue to be emphasized in subsequent years, the report noted that new industrial plants established in the service area during 1958 produced a wide variety of products, including cutting tools and dies for the automotive industry, paper bags, brass fittings, foam rubber materials, carpets, refrigerated soft drink dispensers, staples and stapling machines, fiber glass articles, and meat packing and storage. There was also a new textile research and engineering plant. The long hoped-for diversification, in other words, was fast becoming a reality.

Industrial development in the service area in 1958 included 118 new and 163 major addition to plants. That represented an investment of over $155 million and an estimated payroll of $50 million. While textiles, furniture, and tobacco products—the old standbys—continued to be important, the diversified growth in the area was led by chemical and allied interests, with some $50 million of new investments.[15]

Near the end of 1958, the company introduced a new all-electric rate, which was intended to stimulate the installation of electric heating and heat pumps. In the residential area, the average use per customer was 4,858 kilowatt hours. That was 40 percent greater than the national average, which was about the same in 1958 as had been the 1953 average for Duke Power customers. And the cost to Duke Power residential customers averaged 2 cent per kilowatt hour, which was (and would long remain) about 20 percent less than the national average.

In the matter of electric energy production, of the total 12.5 billion kilowatt hours produced in 1958, 85.3 percent was supplied by steam plants, 13.3 percent by hydro plants, and 1.4 percent purchased from other companies. Substantial reductions in the costs of steam generation had been realized since 1947 by the reduction in coal requirements per kilowatt-hour of production. This reduction was explained by the installation of new and more efficient steam plants and additional units in existing plants. The use of 1.11 pounds of coal per kilowatt hour in 1947 fell to 0.77 of a pound in 1958, which represented a reduction of over 30 percent.

In the personnel area, the 1958 report noted a new emphasis on employee development. As part of a training program to develop qualified personnel, the company again sent selected supervisors to special short courses in utility and general business management and safety at the Georgia School of Technology. In human and industrial relations, a group of employees attended the Southern Industrial Relations Conference at Blue Ridge, North Carolina.

Within the company, training courses were expanded to include training in conference leadership for over 100 selected supervisors and in employee-customer relations for applicable personnel in the general office. The company gave a ten-hour, three-part course in Basics of Supervision for all supervisors in the system. A new management development program for senior executives featured courses taught by professors and other specialists in the new Lake Hickory Training Center.

An extensive, system-wide, two-way employee communication program began in 1958. The basis of the program was monthly discussion periods led by trained conference leaders on the "American Economic System." All employees were given the opportunity at these meetings to discuss the material presented and to direct questions about the material for answer by a committee composed of top management. Finally, the report noted that over 200 employees had enrolled in over 270 job-related college courses under the company's Tuition Refund Program begun in 1957.[16]

Another theme emphasized in the 1958 report as well as in those for subsequent years dealt with safety. The company's Safety Program had helped reduce the injury-frequency rate from 5.88 in 1957 to 4.40 in 1958, a decrease of approximately 25 percent. Accordingly, Duke Power received an award from the Edison Electrical Institute (the industry's trade association) for the reduction in injury frequency and also the award of merit from the National Safety Council. Both the steam and operating departments received awards from the Edison Electrical Institute for operating over 1,000,000 manhours without a disabling injury.[17]

The report also included information on Duke Power's involvement in research on nuclear power, which has been discussed earlier. And, ending on a realistic note, the report pointed out that taxes claimed the largest share of revenue during 1958, amounting to one-fourth of total revenues. About 58 percent of the tax total went to the federal government and 42 percent to state and local governments.[18]

Obviously, such a report as summarized above was of value and interest not only to shareholders but also to journalists, government officials, and ordinary citizens of the Piedmont Carolinas who concerned themselves with economic growth and development. Duke Power was moving to present itself more fully to a larger audience. One reason for this became clear in July 1961, when for the first time common stock in Duke Power (DUK) was traded on the New York Stock Exchange.

The stock had previously been accorded unlisted trading privileges on the American Stock Exchange, but the new listing in 1961 clearly afforded a wider opportunity for trading and a broader market for future company financing—something that grew more important each year.

The Piedmont Carolinas' economy was enjoying an unprecedented boom, and the addition of new generating units at Duke Power's steam plants became a routine affair. Lee Steam Station gained a new unit in 1958, but Allen's unit 3 that went into operation in 1959 stood out: costing some

$40 million, the 275,000-kilowatt generating unit was the largest on the Duke system and one of the largest in service anywhere at the time. When units 4 and 5 were added to the Allen plant in 1960–61, it became the largest investor-owned steam-electric plant in the Southeast and brought Duke's total system nameplate capacity to 3,602 megawatts.[19]

As Duke Power enjoyed a boom, along with its service area, a national business magazine's story about the company in 1962 affords an insight into its growing national reputation. *Forbes* entitled its story "Duke's Legacy" and began with the assertion that "not the least of tobacco king James B. Duke's achievements was Duke Power—the United States' most efficient investor-owned utility." When J. B. Duke died in 1925, the article continued, Duke Power, still largely hydro-powered, reported sales of $20.8 million, generating capacity of 634,029 kilowatts, and net profits of $5.9 million. For 1961, W. B. McGuire happily reported revenues of $178 million, generating capacity (14 percent hydro) of 3.6 million kilowatts, and net profits of $27.2 million. In the previous decade alone, Duke Power's revenue had doubled and its earnings per share shot up 174 percent—a record, according to *Forbes*, that only a couple of fast-growing utilities in Texas and California could match.

Forbes found Duke Power a different kind of utility in other ways. Besides providing water and bus service to a handful of cities in its service area, it operated a yarn mill (acquired because of the mill's hydro site), a wholesale electric equipment firm (Mill-Power Supply), and conducted forestry and land-rental operations. The company's $6 million investment portfolio included a large block of stock in textile producer J. P. Stevens Company and a 12 percent interest in the Piedmont and Northern Railway.

Installing some 2.6 million kilowatts of new steam capacity in just fifteen years after World War II, Duke Power had a "breakneck expansion program" that, *Forbes* noted, was exceeded only by such utility giants as American Electric Power in Ohio and Southern California Edison. On a system basis, Duke Power, according to the Federal Power Commission, ranked as the nation's most efficient investor-owned electric utility.

Compounding its operating advantage, Duke Power since 1945 had lessened its dependence on the textile industry, although the company continued to furnish electricity to about 40 percent of the nation's spindles. By 1961 Duke garnered over 26 percent of its industrial revenues from nontextile producers, as compared with 19 percent a decade earlier. At the same time, Duke's residential consumption, which was 40 percent higher than the national average with rates 20 percent lower, rose to over 25 percent of the load in 1961.

Thus in 1961, *Forbes* concluded, despite a recession, Duke Power set new records for sales and per share earnings for the tenth year in a row. On a 7 percent rise in revenue, net income was up 6.4 percent and per share earnings up 3.2 percent. "I see no reason why," McGuire was quoted as saying, "the next ten years shouldn't be as good as the last ten have been."[20]

McGuire's optimistic forecast turned out to be correct—almost. The glory era for Duke Power would last until 1969, when serious new problems arose. Meantime, expansion and profitability continued to be the hallmarks of most of the 1960s. Despite the massive shift to steam plants, Duke Power had not forgotten hydroelectricity. In 1963 a new hydro plant on the Catawba, Cowans Ford Hydro Station, began operating. Hydro plants are much more easily started and stopped than steam plants, and for that reason, among others, Cowans Ford was intended to help with peak loads, a matter that would become critically important, even urgent, in the 1970s.

While Duke had announced the project in 1957, in the company's report for the following year McGuire explained an important point: "In the opinion of the Company, the construction of the Cowan's Ford development will complete the feasible development of the available head in the stretch of the Catawba-Wateree River above Camden, South Carolina."[21]

In other words, almost sixty years after Dr. W. Gill Wylie and W. S. Lee completed the first hydroelectric plant on the Catawba, Duke Power used the last economically feasible site for hydroelectric production on the river at Cowans Ford.

Much more was involved in the project, however, than one hydro plant. The vast dam (6,846 feet long, 130 feet high) impounded a huge reservoir that the company named Lake Norman in honor of Norman Cocke. Covering 32,500 acres and with a 520-mile shoreline, Lake Norman was the largest lake in North Carolina. While Duke Power was thinking "cooling water for steam plants," tens of thousands of Tar Heels began quickly to think of recreation—swimming, boating, fishing—and lakeside homes.

Duke Power had acquired almost 34,000 acres of the land for what became the Cowans Ford-Lake Norman project back in the 1920s, and the average cost was almost $44.00 per acre. Then from 1957 to 1962, when it was generally known that the development was imminent, the company had to purchase an additional 29,574 acres at an average cost of almost $253.00 per acre.[22]

Some landowners chose to sell entire tracts of land while others chose (wisely, as it turned out) to sell just the portion of land that would be inundated by the proposed lake. Duke Power, at any rate, eventually found itself in the real estate business in a large way.

Owning about half of the lake frontage surrounding Lake Norman, Duke Power gave the state of North Carolina 1,458 acres of wooded land with nearly nine miles of water front for use as a state park. The company also gave almost 110 acres of its Lake Norman land to Davidson College for a water-sports and recreation area. As had long been the practice on many other company-owned lakes, Duke Power provided numerous public-access areas on Lake Norman, with ample parking spaces.

Duke Power from an early date began to give long-term, inexpensive leases on its lakefront lands as a method of fostering good public relations. Following that policy when Lake Norman was filled in 1963, the company began giving a ten-year lease on lakefront lots for $120 a year. When truly difficult financial problems befell Duke Power in the mid-1970s, the management, desperate for cash, began to re-examine the leasing policy and decided to have a system-wide appraisal of all of its lakefront lots, from Lake James up at Bridgewater down to Wateree. The appraisers came up with a figure of about $8,000 a lot, so Duke sold some 6,000 lots at that price, even though some Duke managers considered the price too low. (In the late 1990s, McGuire noted that, "A lot on Lake Norman, a nice lot, is $200,000."[23])

Fortunately for Duke Power, it did not divest itself of all its valuable real estate on Lake Norman. Certain extensive areas that were initially planned to be utilized as sites for steam plants turned out not to be needed for that purpose. Consequently, the company's separate real-estate subsidiary—which originated in 1963 but became Crescent Resources, Inc., in 1989—still had a good bit of extremely valuable real estate to develop.

Lake Norman has clearly played a major role not only in Duke Power's long involvement with the matter of harnessing the Catawba River but also in the lives of hundreds of thousands of Tar Heels in the Piedmont. To illustrate just one small segment of the careful planning and thinking that went into the vast project, Bill Lee later explained about the land clearing that the company had to undertake for the lake. "On Lake Norman, we had two principles that were conflicting," Lee noted. One involved the propagation of fish. Leaving the trees in the lower part of the lake would have provided hiding places for small fish and increased the fish population faster. "It would also," he continued, "provide good 'hot holes' for fishmen seeking the large predator fish." That would obviously have been applauded by many devotees of fishing, especially in the first twenty-five to thirty years before the trees fully decayed.

The other principle involved dissolved oxygen in the waters of Lake Norman. The decaying trees would have reduced the oxygen dissolved in the water and made the lower levels oxygen deficient. Not only would that not be helpful to fish growth, but, more importantly, the reduced oxygen in Lake Norman water would reduce its ability to oxidize other contaminants as well as to provide oxygen-rich water released downstream for the major water supply intakes of Charlotte and Gastonia on Duke's Mountain Island reservoir. "We decided that water quality had to prevail," Lee declared, "and we scraped Lake Norman clean, cutting off the stumps at the ground line with specially prepared bulldozers." Of course, he added, "merchantable timber was removed before the land clearing was done."[24]

Clearly, Lake Norman served many purposes in addition to the power-generating needs of Duke Power. Intricately involved with the economic and recreational life of its service area, the company in 1964 moved to strengthen its identification with the region. J. B. Duke had incorporated the Duke Power Company under the laws of New Jersey. In 1964, however, Duke Power incorporated under the laws of North Carolina because all of its customers, some 60 percent of its shareholders, and virtually all of its physical properties were located in the Carolinas, with the majority in North Carolina. Moreover, the change much simplified the matter of issuance of new securities.

At the same time that Duke Power moved to tie itself more closely to its service area, it also acted to spread the word nationally about the economic attractiveness of the Piedmont Carolinas. The company took full-page advertisements that touted the region in such publications as *Business Week*, *Nations Business*, *Fortune*, the *Wall Street Journal*, and other key media. Moving to strengthen ties with the national financial community, in 1963 Duke Power invited nearly a hundred investment bankers and security analysts and dealers from the New York area and from cities in the Carolinas to Charlotte. There they met with the Duke management team, some members of the North Carolina Utilities Commission, and many of the leading banking, industrial, and business leaders in the area. The visitors were also taken on a tour to see first hand the existing industrial development and the future potential of the Duke Power system. The visitor's comments suggested that they were impressed not only by the on-scene operations of Duke Power but also "by the economic vigor of the Piedmont section of North and South Carolina."[25]

Reporting to the shareholders about 1965, McGuire declared that it was "by far the biggest year of industrial expansion in the Company's history." New plants and plant expansions announced for the Duke service area represented an investment of over $561 million, which was a 43.6 percent increase over 1964 investment. That 1965 broke many records for the company—in electric revenue, energy sales, average kilowatt-hour usage, peak load, generating capacity, the number of all-electric homes—was hardly surprising in such a vibrant economy.[26]

Icing the cake for Duke Power customers was the fact that the company announced late in 1965 its sixth rate reduction since 1960. The reductions meant savings for Duke customers of over $9 million per year. (This string of rate cuts in the 1960s would heighten the shock of rate increases that became essential in the inflation-plagued 1970s.)

Carrying out the company's well-publicized plan for the utilization of Lake Norman's ample cooling water, Duke Power built the Marshall Steam Station, and the 385,000-kilowatt unit 1 of the plant began operating in March 1965. Named for Edward C. Marshall, the plant utilized coal. As to whether the Marshall plant's next two generating units, each rated at 660,000 kilowatts, would be nuclear or coal fired became a subject of careful study and debate within the company. When negotiations for the low-sulphur coal that Duke utilized were successful and the railroad companies agreed to economically competitive rates, Duke Power decided in favor of coal for the additional units at Marshall.

Since freight rates were a significant component of the overall cost of coal, Duke Power succeeded in getting the rail companies with which it dealt to utilize a new "unit train" policy. Under this plan, specific railroad equipment was assigned to scheduled runs between the mines in Kentucky and West Virginia and Duke's steam plants. Given high priority by the railroads, the unit trains maintained extremely close schedules. Moreover, the coal-loading operations were drastically changed through the use of a facility known as a transloader. Coal gathered from various mines was brought to the transloader, where it was blended to get the desired heat-content mix. The coal was then either stored in silos or quick-loaded into unit trains. At the power stations the use of high-capacity unloading equipment assured a quick turn around for the trains, and the resulting high utilization of the coal cars enabled the railroads to offer a lower freight rate than had been possible for separate carload shipments previously used.[27] Thus, Marshall Steam Station missed out on having Duke Power's first nuclear-fired generating units, though that would soon be announced for another location. Marshall did have the distinction of being the first computerized coal-burning steam plant on the system.

Duke's construction program was about to accelerate and expand significantly, and as a result of the do-it-yourself construction policy the number of employees would also rise. At the end of 1965, however, there were 6,235 men and women on the Duke payroll. Among them, 58 percent had more than ten years of service and 26 percent more than twenty years. In other words, it was clearly not just the company's top and mid-level managers who chose to make a long-term commitment to Duke Power. The richly experienced Duke engineers and workers were about to help the company significantly outperform other electric utilities in the nation when it came to the exacting and expensive business of building nuclear plants.

Early in 1965 Bill McGuire announced a mammoth and trail-blazing new venture for Duke Power, the Keowee-Toxaway Project. Named for two rivers in the ruggedly mountainous corner of South Carolina and a few miles northwest of Clemson, among other things, it would involve two firsts for the company: a nuclear plant and a pumped-storage hydro plant.

Although Duke engineers had spotted the potential of the area as early as 1916 and the company had purchased some land for dam sites, Duke Power formed a new subsidiary in 1963, the South Carolina Land and Timber Company (which later became Crescent Resources), to buy a big block of 31,113 acres for the projected multipurpose Keowee-Toxaway development. The average cost was $166.55 per acre. In 1965, Duke found it necessary to publicly announce the project, and the additional 21,526 acres that had to be purchased cost an average of $449.10 per acre. In other words, after the public announcement the price of the land almost tripled. No wonder the company, following J. B. Duke's consistent policy, tried, when it could, to buy as much of the land needed for a development before it was publicly announced.[28]

No sooner had Duke Power applied to the Federal Power Commission early in 1965 for permission to build the first phase of the Keowee-Toxaway project than opposition to it sprang up. Before that is explained, however, perhaps certain background information might be helpful.

The 1960s, with the administrations of Presidents John F. Kennedy and then Lyndon B. Johnson, produced majorities in Congress that displayed great faith in the power and ability of the Federal government to solve many of the nation's social and economic problems, even as the Cold War intensified and became a hot war in Vietnam. President Johnson especially aspired to expand many of the New Deal's programs and even to move into areas, such as medical care for the aged (65 and older) and the indigent, that Presidents Franklin D. Roosevelt and Harry Truman had not been able to reach. Majorities in Congress, as in the nation, supported Johnson's ambitious programs.

The issue of public-versus-private electric power continued to be very much alive. Not only had the Tennessee Valley Authority kept growing, but another child of the New Deal, the Rural Electric Administration and its

REA cooperatives flourished with the assistance of the Federal government. The electric cooperatives had moved, in some areas, into power generation, and they increasingly served large numbers of non-farm customers. Moreover, the REA cooperatives included enough members scattered around many areas of the nation to have considerable political clout.

The investor-owned electric utilities were no happier in the 1960s about these government-assisted competitors than they had been back in the 1930s. Duke Power's Bill McGuire, in particular, was a feisty, outspoken critic of Federal favors for selected groups of power consumers at the expense of the nation's tax payers. Unfortunately for McGuire, no one pointed out to him some famous words of President Andrew Jackson back in 1832:

> There are no necessary evils in government. Its evils exist only in its abuses. If it would confine itself to equal protection, and, as Heaven does its rains, shower its favors alike on the high and the low, the rich and the poor, it would be an unqualified blessing.[29]

Even without Jackson's eloquent call for "equal protection," McGuire had no trouble presenting his case. With the Cold War intensifying in the early 1960s, the Kennedy administration placed top priority on maintaining the nation's defensive strength and its role as leader of the free world. "In maintaining the economic health necessary to both these objectives," McGuire noted, "a financially strong utility industry is essential, since our industry is the largest in invested capital, the largest in annual construction of new plant, and the largest single taxpayer of all U. S. industry." It had also placed the United States "far ahead of any other nation in total power capacity and production of electricity per capita."

The leadership of the United States in the production and use of electricity, McGuire continued, had been accomplished by investor-owned, regulated, tax-paying electric companies that supplied 80 percent of the electric power used in the nation. Moreover, these companies had plans that would meet the expanding demand for electricity.

Despite the fine record of service by the utility companies and the heavy tax load carried by their customers (25 cent of each dollar paid for electric service went for taxes), the Kennedy administration was pushing to expand tax-subsidized government production and supply of electricity. The Rural Electrification Administration had asked for a 40 percent larger appropriation to make 2 percent loans to REA cooperatives so that they could build generating plants and transmission lines. The Department of Interior also sought funds for the construction of extensive transmission lines and for the building of additional hydroelectric plants.

McGuire argued that the Federal government should be decreasing rather than increasing expenditures in these area for three reasons: 1) They were unnecessary because investor-owned companies could and would meet all requirements for electric service. 2) Increased Federal expenditures would further burden already hardpressed taxpayers of the nation. And 3) the increased expenditures were unfair because they resulted in government-subsidized competition with the tax-paying, investor-owned electric utility industry.[30]

McGuire conceded that in its original purpose to help make electricity available to the farmers in the 1930s and 1940s the REA had been highly useful and helpful. That purpose, however, had been accomplished, and by the 1960s the tax-exempt REA cooperatives were making more than half of their sales to non-farm users, and five out of six new REA customers were non-farm. To heighten the irony, McGuire noted that the average cost per kilowatt hour to REA cooperatives of power from REA-financed generation was higher than the average cost of power bought by REA cooperatives from tax-paying companies. Yet the Federal government continued to make low-cost, taxpayer-subsidized loans for REA generating plants.[31]

In short, McGuire charged that "cheap government power" was cheap only to the preferred customers—electric cooperatives and municipalities—not to the taxpayers. The government did not charge the true economic cost of the electricity, because the government (and its agencies) did not pay taxes or fair interest charges. "Government in the power business," McGuire concluded, "favors some electric users at the expense of the great majority whose electric bills include full taxes."[32]

McGuire's arguments would probably have received a more sympathetic hearing from a larger number of Americans in the 1990s than they did in the 1960s. The 1960s, like the 1930s, were characterized by a mood or sentiment that cheered on—or at least accepted—an ever-widening role for the Federal government. By the later decades of the century, however, many more people had converted to free-market views as expressed by one of the architects of Hong Kong's famous competitive, free-market economy. Fighting off a proposal that lower water rates should be funded out of the government purse (i.e., by taxpayers), Sir John Cowperthwaite argued: "I see no reason why someone who is content with a cold shower should subsidize someone who is able to luxuriate in a deep, hot bath, or why someone who waters a few plants on the window-sill should subsidize someone who waters his extensive lawn." As for government's entering into business, he suggested, "One trouble is that when government gets into a business it tends to make it uneconomic for anyone else."[33] Bill McGuire, an outspoken foe of government's steady intrusion into power production, would have applauded.

Against this background of the ongoing conflict between the investor-owned utilities and the champions of public power, various aspects of the Keowee-Toxaway story became more understandable. The truth was that Duke Power and the public-power forces were engaged in a contest about the development of the upper reaches of the Savannah River and its tributary system in the mountainous corners of the Carolinas.

Northwest of Augusta, Georgia, at elevation 330 feet on the Savannah River, the Federal government had built the Clark Hill dam and hydroelectric plant, which began selling power, with "preferred customers" having first priority, in the mid-1960s. Further upstream at Trotters Shoal, elevation 475 feet, the Federal government proposed to build another dam and plant. This project, however, ran into a snag: at the northern end of the proposed Trotter Shoals reservoir, Duke Power had long owned a property known as Middleton Shoals and proposed to build a diversion dam with a relatively small reservoir that could be used to supply cooling water for a large steam plant. Having failed in earlier attempts to get a bill through Congress authorizing the construction of the dam, Duke Power finally succeeded in gaining passage of the measure in October 1966.

Beyond Middleton Shoals and further up the Savannah River, the Federal government had the Hartwell dam and generating plant at elevation 660 feet. Then beyond that and getting into the tributary system of the Savannah, Duke Power in 1965 announced plans for its vast development known as Keowee-Toxaway. The proposed Keowee Lake and hydroelectric plant would be at elevation 800 feet and the proposed Lake Jocasee and pumped-storage hydro plant, some 11 miles upstream, would be at elevation 1,000 feet. In other words, if one thinks of the upper reaches of the Savannah River and its tributary system as a sort of staircase of power sites, both developed and potential, then Duke Power and the Federal government occupied virtually alternating risers or levels.

Duke's Keowee-Toxaway plans had strong political support in South Carolina, and that would play a large role in the final outcome. Congressman William Jennings Bryan Dorn of South Carolina typified the reaction when he hailed the proposed Duke project as the "greatest single industrial announcement in the history of South Carolina . . . an industry the magnitude of which is fantastic and almost incomprehensible."[34] The Keowee development involved dams on the Keowee and Little Rivers; the resulting reservoirs, covering over 18,500 acres, and with a shoreline of 300 miles, would be connected by a canal to form Lake Keowee. The hydro generation at Keowee dam would be about 140,000 kilowatts.

Then another dam and hydroelectric plant would be located higher up the Keowee River, near Jocassee and near where the Whitewater and Toxaway Rivers join. The Jocassee development was planned for 160,000 kilowatts of hydroelectric capacity and ultimately 450,000 kilowatts of pumped storage, a total installation of 610,000 kilowatts. (In a pumped -storage facility, off-peak electricity from base-load, constantly-operated steam plants is used to pump water from a lower reservoir—in this case, Lake Keowee—to an upper reservoir—Lake Jocassee. That water then becomes available for hydroelectric generation for use during peak-period demand. While electricity can not be stored economically, water can be—and in a pumped-storage system, it can also be re-used repeatedly. The matter of meeting peak-period demand would become increasingly critical from the 1970s onward. Since the Duke Power management and engineers did not know this in 1965, they showed considerable foresight—and perhaps some good luck—in anticipating the problem.)

In announcing Keowee-Toxaway, Duke Power officials noted that several other sites on the North and South Carolina tributaries of the Keowee were contemplated for future development. In addition to the hydropower developed, the lakes could provide sites for steam generating stations totalling 7 million kilowatts. (For comparative purposes, Plant Allen, then the largest on the system, was rated at 1.25 million kilowatts.)

In making the original announcement of Keowee-Toxaway, McGuire noted that Duke Power had requested Congressional authorization to build a steam generating station in Anderson County, South Carolina, at Middleton Shoals on the Savannah River. "If that authorization is received," he explained, "that will be the next steam plant we will start." If the request should not be approved, however, one of the Keowee steam plants could replace it.[35] (Although Duke finally received the authorization to build at Middleton Shoals, the company chose instead to build Oconee on Lake Keowee.)

Along with power generation at Keowee-Toxaway, there were to be a wide variety of recreational attractions. Duke Power had acquired more than 100,000 acres in the area surrounding the lakes and extending up the watersheds of the Whitewater, Toxaway, Horsepasture, and Thompson Rivers into North Carolina. Since protection of the watersheds was vital to the operation of the power projects, the area was to be kept under scientific forest management. That would continue to provide timber for local mills and jobs for people working in forest industries.

The forest management would also provide a great variety of outdoor recreational opportunities, including hiking and riding trails, hunting, fishing, and camping. All these, together with the elaborately planned Duke Power Visitors Center (later called the Oconee World of Energy) on Lake Keowee, would attract tourists and new businesses.

Duke Power had leased 68,000 acres to the South Carolina Wildlife Resources Department for a game refuge. It would be known as the Horsepasture Game Management Area, so named for the Horsepasture River, a wild-water trout stream that flowed from North Carolina's Transylvania County into Oconee County, South Carolina. As at so many other Duke Power lakes, the company planned to develop free access areas on the Keowee lakes. Sites were to be leased to commercial operators to develop campgrounds, picnic areas, golf courses, club houses, marinas, and tourist accommodations. Certain parts of the area, however, such as Whitewater Falls, had high scenic value and would be preserved. If the state wished to make these scenic areas available by highway, Duke Power expressed its willingness to cooperate.

Finally, Duke's practice was to allow towns and industries to draw their water supplies from its reservoirs without charge. Fifteen towns and cities already got their water from company reservoirs, as did the Chester (South Carolina) Metropolitan Water District.

The Federal Power Commission had encouraged the electric utilities to engage in long-range planning. McGuire asserted his belief that with the Keowee-Toxaway development Duke Power had done precisely what the Commission had wanted.[36]

Nevertheless, there were certain champions of public power who hoped to pull the plug on Duke Power's big project. A group of REA electric cooperatives in Georgia and the two Carolinas petitioned the Federal Power Commission for the right to intervene in the matter of Duke Power's application for a license to begin work on Keowee-Toxaway. Having organized in 1960 to obtain power from the Federal government's Hartwell plant, the committee also worked to have Congress authorize funds for Federal development of the entire Savannah River basin, including its tributaries.[37]

From the power company's standpoint, such intervention as that of the Tri-State Power Committee posed a problem. There were numerous bureaucratic hurdles that had to be jumped in the process of securing a license to build a power plant. Even if the intervenors did not ultimately succeed, as they did not in this case, they could frustrate carefully worked-out construction schedules and thereby add to the cost of a project. (This would be one of many factors to complicate and increase the cost of nuclear plant construction in the 1970s and 1980s.)

> "We are surprised and greatly disappointed over this action by the electric co-ops," Bill McGuire promptly announced. The intervenors, if successful, would destroy a major taxpaying industrial development so that sales of power from a government project could be made to tax-subsidized "preferred customers."[38]

Various South Carolina political leaders were even more outraged and outspoken about the cooperatives' action than McGuire had been. One state senator from Pickens County in the area of the project declared that if it should not be built, "We would consider it the greatest loss in the industrial history of the state." Another state senator, pointing to the huge state and local tax bill that Duke Power would pay for Keowee-Toxaway, insisted that the Tri-State Committee certainly did not speak for the cooperative in his county. The senator from Abbeville County insisted that "free enterprise" should be given the chance "to develop all over the state" and that if the co-operatives were set to fight private power projects, South Carolina was "going to be handicapped and stymied" in pursuit of industrial development.[39]

So widespread and loud was the support in South Carolina for Duke Power and its proposed project that the South Carolina cooperatives were forced to withdraw from the Tri-State Committee, leaving their Georgia and North Carolina associates in a bi-state coalition. But another intervenor cropped up from a surprising quarter.

At the last possible moment for such action, none other than Secretary of Interior Stewart L. Udall petitioned to intervene in the Keowee-Toxaway matter. Suggesting, among other things, that the proposed power plants were not really needed and that Duke Power could always buy power from Federal projects on the Savannah River, Udall's action provoked such a firestorm of opposition in South Carolina that he withdrew his petition a few weeks later. (A cartoon in the *Charlotte Observer* showed a discomforted-looking Secretary Udall seated in an airliner with thunderheads and lightning flashes visible through the plane's windows. To the passenger seated next to him, Udall says, "Is the air always this bumpy over South Carolina?"[40])

Bill McGuire was not about to let Udall go quietly into the night. The Interior secretary's proposed intervention, McGuire charged, had "brought bureaucratic empire building into raw and open conflict with the enterprise system upon which our economy is built." It had been "quite bewildering to all of us," McGuire continued, "to see an agency of the same Federal Government which advocated and obtained passage from Congress of the Aid to Appalachia Bill register official opposition to our request for authority to invest hundreds of millions of dollars of private capital in the Appalachian Region in order to meet the power needs of our customers."

As for Udall's suggestion that Duke Power would not need the generating capacity proposed to develop at Keowee-Toxaway and that Duke could buy Federal power instead, McGuire fired back: "To the best of my knowledge this is the first time that the Secretary of the Interior or any other officer of the Federal Government has ever suggested that he knew more about our future power requirements than we did." It was also the first time, McGuire continued, that any Federal agency had taken the position that "we should not build our own generating plants, but should buy our hydroelectric power from Federal Government plants." McGuire reiterated that not only did "preference customers" have first priority for Federal power, but also that such contracts as Duke did have to buy certain blocks of Federal power (as at the Hartwell plant) were for five years only. "Obviously, we cannot be dependent upon such short term sources of electricity for supplying the needs of the rapidly developing area served by the company," he declared.

"Mr. Udall predicated his intervention upon forcing the will of the Federal Government upon the state of South Carolina," McGuire concluded. If the Secretary of the Interior had accomplished his stated objective, "not only would the company's ability to plan and prepare for the future have been thwarted and the area subjected to further Federal direction and control, but in addition the local and state governments would have been deprived of the tax support which will be provided by the Keowee-Toxaway Project. . . ."[41]

With one intervenor out of the way, Duke Power still faced the problem of the Bi-State Committee's intervention. Fearing that hearings before the Federal Power Commission might drag on indefinitely, McGuire and his associates were eager to avoid that. With the valuable assistance of Congressman Bryan Dorn of South Carolina, Duke Power cut a deal. At a long, occasionally heated meeting in Dorn's office in Washington, officials of the Bi-State Committee agreed to withdraw their intervention in the Keowee-Toxaway matter in exchange for a promise that Duke Power and the South Carolina congressional delegation would no longer oppose Federal construction of the dam at Trotters Shoal on the Savannah River. (It ultimately became the Richard Russell Dam.) The REA cooperatives also agreed to withdraw their opposition to a diversion dam or a possible Duke plant at Middleton shoals on the Savannah.[42]

Finally, in September 1966, Duke Power gained its license to proceed with the first phase of the giant Keowee-Toxaway project. Even before that, however, the company announced in July 1966, that the first steam plant to be built on Lake Keowee would use nuclear fuel, and in early December 1966, Duke Power filed an application with the Atomic Energy Commission for a construction permit for what would be the Oconee Nuclear Station, the company's first such facility. Although Duke first talked of two generating units at Oconee, it subsequently planned for a third unit, giving the plant a net capability of 2.6 million kilowatts and making it the largest generating plant announced up to that time in the United States. Including the third unit at Oconee, the total cost of the whole Keowee-Toxaway development was now projected to be almost $700 million. By the late 1960s, in other words, Duke Power's need for capital would be enormous—and unforeseen trouble, stemming partly from increasing inflation, lay ahead.

More immediately, there was also opposition to the company's application for the Oconee license. The opposition came not from militant antinuclear groups (they would appear later), but from eleven North Carolina municipalities that purchased wholesale power from Duke but operated their own distribution systems. The municipal protestors argued that the licenses applied for by Duke Power were not authorized by the Atomic Energy Act of 1954, and that an unconditional license would appear to violate or tend to violate antitrust laws and restrain free competition. The cities asked that they be permitted to share in the cost of constructing the Oconee Nuclear Station so that they could buy electricity at cost. Finally, they argued that nuclear power had been developed at government expense (i.e., by the tax payers), so the results should be enjoyed by the public without discrimination.

Denying any merit to these arguments, B. B. Parker, then Duke's executive vice president, explained first that the company was not surprised by the action. He noted that the eleven municipalities had hired the same lawyers that Fayetteville, North Carolina, employed to win a rate reduction from Carolina Power and Light. Protesting on antitrust grounds against CP & L's nuclear plant being built near Hartsville, South Carolina, Fayetteville abandoned its suit only when CP & L granted a rate reduction.

"Duke has no intention of granting a wholesale rate reduction," Parker declared, "as our wholesale rate is lower than any other in the Carolinas. . . ." He argued that the attempt to delay construction of a needed power facility was "outrageous and jeopardizes the future power supply of the entire Piedmont Carolinas." As for the claim that the eleven cities should "own" a part of Oconee, Parker argued that if that right were granted to them, it would have to be granted to "every one of Duke's 860,000 customers."

As for the notion that atomic power had been developed solely by the Federal government, Parker pointed out that investor-owned utilities shared with the Atomic Energy Commission in the costs of the experimental Parr Nuclear Station. Furthermore, governmental research in the nuclear area had been paid for by Federal taxpayers. The municipal systems, however, paid no Federal taxes—in fact, no taxes at all—except that amount paid indirectly through the purchase of wholesale power from Duke. "And now," Parker added, "these towns were trying to escape even those taxes." The municipal officials wanted Federal benefits but were unwilling to pay Federal taxes—"and they talk about discrimination!" Parker emphasized that the Oconee plant would produce annually about $11 million in Federal taxes and $9 million in state and local taxes.[43]

Parker's arguments apparently made sense to the Atomic Energy Commission, for in November 1967, it rejected the application of the eleven North Carolina cities to acquire an ownership interest in Oconee Nuclear Station. (A decade or so later, in vastly changed circumstances, Duke Power would reverse its position about municipalities owning a share of its nuclear plants.)

With intervenors and protestors out of the way, work on Keowee-Toxaway proceeded steadily. While that went on, Duke Power fed the Piedmont Carolinas' voracious appetite for electricity by steadily adding new generating units at its Lee, Dan River, Marshall, Riverbend and Buck plants—between December 1967, and July 1970, nineteen such units were added.

Keowee Hydro Station began commercial operation in April 1971, with two units at 70,000 kilowatts each. Then in July 1973, Oconee Nuclear Station's unit 1 began to produce electricity for sale. (Units 2 and 3 did the same in 1974.) Thanks to the company's do-it-yourself construction policy, Duke completed Oconee's three units in seven years at a total cost of $500 million. That averaged out at $194 per kilowatt, which was less than half of the industry average at that time. At completion, Oconee's reactor buildings were the largest pre-stressed, post-tensioned concrete buildings in the nation.[44]

Marking the completion of at least the first phase of the Keowee-Toxaway development, Jocassee Hydro Station's first two units, the company's first pumped-storage facility, began commercial operation late in 1973. (Units 3 and 4 followed in 1975.) The construction of Jocassee Dam had required more than 11 million cubic yards of earth and rock. By the time of the impressive dedication ceremony for Keowee-Toxaway in October, 1973—nearly nine years after the first announcement of it—Duke Power had ample reason to be proud. The previous year it had received its first Edison Award, the industry's most coveted recognition, from the Edison Electric Institute for the planning and executing of Keowee-Toxaway. The citation, containing understandably sweet words to Duke Power managers and employees, read: "For its outstanding engineering accomplishments in the integrated hydro-thermal development of the Keowee-Toxaway Project, fully utilizing the area and its natural resources for electric generation and at the same time protecting and enhancing the environment of the Keowee Valley, Duke Power is declared the recipient of the Edison Award for 1972."[45]

As if that were not enough, the American Society of Civil Engineers selected the Keowee-Toxaway project as the nation's outstanding achievement in civil engineering for 1975. Some of the other major projects in competition for that year's award were the San Francisco Bay Area Rapid Transit, the Dallas-Fort Worth Airport, and the Chesapeake Bay Bridge.[46]

The Visitors Center at Keowee-Toxaway, opened in July, 1969, and immediately became one of the top tourist attractions in South Carolina. Telling and showing the complete "Story of Energy" and offering magnificent views of the Lake Keowee area, the Center drew a greater attendance than any other such electric industry facility in the nation. The South Carolina Chamber of Commerce picked the Center as the state's top tourist attraction in 1972.

Duke Power management, always proudly conscious of the company's history and thoughtful of those who had played a role in it, arranged for one special visitor to see Keowee-Toxaway in 1972. The company had a helicopter fly Richard Pfaehler there on his 90th birthday. The German-born and educated hydraulic engineer had taken a job with the Southern Power Company in 1908, two years after emigrating to the United States. The four-man engineering department then consisted of William S. Lee as chief engineer, C. A. Mays as design engineer, Pfaehler as draftsman and detailer, and Charlie Torrence as blueprint and "handy boy."

Pfaehler had not only worked on J. B. Duke's hydro project in Canada—Duke was alleged to have teased him about being "a derned Prussian"—but he also helped interest and educate Duke about the whole matter of regulating the flow of the Catawba River.

Pfaehler spotted the potential of the Keowee-Toxaway area at an early date and explained in 1972: "I tried to call their attention to it and they bought a whole lot of land there according to my recommendation." After the flood in 1916, the question became whether it was better to try to regulate the Catawba or utilize the high-head sites in the mountains. Since the latter would require "a lot of big tunnels and pipelines" and since the company already had power stations on the Catawba, the engineers and J. B. Duke agreed that they "should regulate the river and forget about the other for a while." The "while" turned out to be about a half century, but, Pfaehler stated, "I always knew that country up there was one of the nicest sections...."[47] Photographs of Pfaehler and Bill Lee added another nice human dimension to the story.

In addition to the two major awards for Keowee-Toxaway, Duke Power received various other plaudits in the early 1970s: *Financial World* magazine picked the company's *Annual Report* for 1972 as the best among the largest utilities (Duke was then the 6th largest and its *Annual Report* had won the same recognition earlier); Bill Lee, senior vice president of engineering and construction, received the George Westinghouse Gold Medal, the top award given annually by the American Society of Mechanical Engineers; the company received the South Carolina Wildlife Confederation's award as "Water Conservationists of the Year" and the North Carolina Wildlife Federation's Forestry Conservation Award for 1972; and for the seventh straight year (and for many years following) the Edison Electrical Institute named Duke's Marshall Steam Station as the nation's most efficient.

All of that was nice, and it was pleasant news for Duke Power, of course. Unfortunately, however, awards do not put meat on the table, that is, they do not produce cash. And the sad fact was that, beginning in 1969, the company began to get into serious financial trouble. The "glory years" were over, and what Carl Horn later described as "the most critical, frustrating time we ever faced at Duke Power" was about to begin.

CHAPTER V

A TROUBLED TIME FOR DUKE POWER, 1969–1975

The year 1969 saw the beginning of serious trouble not only for Duke Power but also for the Duke Endowment, which had owned since 1924 a controlling interest in the company.When George G. Allen died in 1960, Thomas L. Perkins, lawyer and son of the late William R. Perkins, soon succeeded him as chairman of the Duke Endowment. Then in 1961 the younger Perkins also became chairman of Duke Power's board of directors. In other words, the close, interlocking ties that J. B. Duke had wanted between the power company and the Endowment continued to be very much a reality, and numerous other trustees of the Endowment served as officers and directors of Duke Power.

One of the results of the Tax Reform Act of 1969, however, was to signal the beginning of the end of those close ties. Criticism of certain policies of some of the tax-free philanthropic foundations increased throughout the 1960s. None were more persistent and outspoken in their attacks on the foundations than Congressman Wright Patman, Democrat from Texas. Some of his more extreme ideas—such as a time limit on the life of foundations and substantial taxation of them (so that Congress, rather than private or voluntary organizations, could address more of society's problems and needs)—were ultimately rejected or modified. Patman's efforts, however, finally culminated in the Tax Reform Act of 1969.[1]

Among other things, the Tax Reform Act virtually outlawed close interlocking arrangements between tax-free foundations and profit-seeking businesses, arrangements that had been commonplace in the first half of the century. As of 1957, for example, the Endowment owned over 57 percent of Duke Power's common stock and 76 percent of its preferred shares. The Tax Reform Act, however, mandated that, in order to avoid draconian penalties, the Endowment had to refrain from buying any additional common stock of the Duke Power Company and, within a ten-year period, manage to get its holding of the stock to be less than 25 percent of the total of Duke Power's outstanding common stock. In other words, the Tax Reform Act spelled the doom of J. B. Duke's planned economic base of his perpetual philanthropy in the Carolinas, for his Grand Design had involved having Duke Power be the economic engine that pulled the Duke Endowment. Or, to put the matter another way, J. B. Duke meant for the Endowment not only to have a controlling interest in the power company, but also for the Endowment's trustees to have an active, on-going voice in the management of the company.

The historical record and the testimony of veteran, now-retired officers in Duke Power suggest that J. B. Duke's plan had worked well for the company.[2] Endowment leaders like George Allen and W. R. Perkins and his son Tom Perkins (as well as several others) had shown a lively, steady interest in Duke Power but had never thrown their weight around in ways that might have been harmful to it. The same was not true in the case of the Endowment leaders and Duke University, for down to the 1970s, Allen, the two Perkinses, and a few other Endowment trustees used their power and financial clout in ways that were sometimes harmful to Duke University's best interests. Perhaps because the Endowment leaders were, above all, businessmen, they could respect the integrity and autonomy of a profit-seeking business more easily than in the case of a non-profit and always needy university.

At any rate, the Tax Reform Act of 1969 put the Duke Endowment in a dilemma as far as its relationship with Duke Power was concerned. In the years immediately after 1969, the power company's capital needs were so great that it had to sell an enormous amount of common stock. Simply by *not* buying any of that stock, the Endowment soon saw its share of Duke Power's common stock fall well below the mandated 25 percent level. As for the restrictions that J. B. Duke's indenture had placed on the Endowment's investments (which had to be in Duke Power securities or certain classes of government bonds), the Endowment in 1971 finally gained permission from the North Carolina Supreme Court to begin to diversify its investments, that is to ignore the restrictions that J. B. Duke's indenture had placed on the Endowment's investments.[3]

Exactly how the Duke Endowment's long-standing ownership of a controlling interest in Duke Power affected the company's operation is a difficult matter to assess. As indicated above, the Endowment leaders certainly did not harm or hamper Duke Power. But did the relationship between the two entities have any positive benefits for Duke Power? Looking back from

1997, when the close ties between Duke Power and the Endowment had ceased to exist, several veteran, retired officers of the company asserted their belief that there had been a number of advantages for Duke Power in the relationship.

For example, John Hicks, senior vice president for public affairs when he retired in 1988 and a thirty-year employee of Duke Power, suggested that decision-making had been "very easy" for Duke Power. With so much of the "financial wherewithal" coming from the Endowment and with the overlap of officers and trustees, "We could make decisions in a hurry." That was one reason, Hicks argued, why Duke Power was "so darn good."[4]

Douglas W. Booth, president and chief operating officer of Duke Power when he retired in 1989 and an employee for thirty-seven years, came at the matter another way. "Certainly from a public relations standpoint because of the good works that the Endowment did," Booth suggested, "... that stood the power company in good stead in many instances of problems that came along." Furthermore, Booth believed that the relationship between the Endowment and Duke Power "and the good works of the Endowment made it easy to recruit good people to [work for] the power company."[5]

Bill McGuire, former president and chief executive officer and a forty-two-year veteran with the company, agreed that Duke Power had gained "credit" among especially the leaders in the Carolinas because of its relationship with the Endowment. Yet McGuire was quick to add: "My philosophy was that the Duke Power Company had to stand on its own feet and make its own way and justify itself to its customers, to the public, and its stockholders without any reliance on the Duke Endowment." If the Endowment's activities benefitted the power company, "that was fine, but the power company had no business trying to lean back on that to support any action" it might have taken.

Moreover, McGuire well remembered the 1930s and noted that the Endowment's vehement, long-fought battle against the New Deal's power project in Greenwood County, South Carolina, resulted from "a decision made in New York." The Endowment and Duke Power people in Charlotte "were not of the opinion that that was the thing to do," McGuire declared, but George G. Allen and W. R. Perkins thought otherwise—and they were then the dominant figures in all three of J. B. Duke's institutional legacies.[6]

How much, if any, the company's tie with the Endowment influenced the thinking of Duke Power employees below the management level is not known and can only be the subject of speculation. Throughout much of Duke Power's history, however, there does seem to have been a certain abiding *esprit de corps* and, among many employees, a certain loyalty to the company. Much of that probably derived from the singular autonomy that the company enjoyed from the first, for, unlike so many electric utilities in the South and elsewhere, Duke Power never had to kowtow to a distant, northern holding company. Also, the company's "do it yourself" construction policy after 1924 became a source of pride and satisfaction to many employees besides managers.

But did the fact that the Endowment helped to build so many of the community hospitals scattered across the Carolinas enter the minds of rank-and-file Duke employees? Possibly. Higher education as carried out at Davidson, Furman, Johnson C. Smith, and Duke University may or may not have loomed large for some Duke Power employees. Those that were Methodists, however, could hardly escape being aware that the Endowment significantly assisted that church in North Carolina. The Endowment's support for orphans and childcare in the Carolinas was perhaps not as visible as some of the other philanthropy, but many thousands of Carolinians considered the matter socially urgent and important. In short, Duke Power's ties with the Endowment may well have permeated thinking and affected performances below the management level.

The truth was, however, that the power company's troubles that began in 1969 had little or nothing to do with the Endowment. The inflation that began that year to ratchet upward was the nemesis of Duke Power, along with virtually all other electric utilities in the nation. When the price of coal, especially the low-sulphur coal that Duke Power had long used, began to climb relentlessly upward in 1969, the company got caught in a classic economic squeeze. As a regulated monopoly, it was not free simply to do what most businesses do: pass rising costs on to consumers through higher prices.

The rapidly rising cost of capital, that is, the interest that had to be paid on money borrowed through the sale of bonds, further complicated matters. While there was nothing Duke Power could do about the escalating cost of capital, the company did attempt, in vain initially, to do something about the soaring expense of its primary fuel, coal.

The Environmental Protection Act that Congress passed and President Richard Nixon signed into law in 1969 had profound implications for many aspects of American life and business. Among them, especially for the electric utility industry, was the premium that the new legislation placed on coal that produced less of the pollutant sulphur dioxide when burned than did the nation's vast deposits of high-sulphur coal.

In 1968 the cost of coal for Duke Power averaged 28 cents per million BTU (British Thermal Unit). By March 1970, it had risen to 35 cents—a 25 percent increase in less than two years. Predictions by various knowledgeable people were that further rises in coal prices were bound to come, and of course they did. At that time, 31 cents out of every dollar of Duke's gross revenue went to pay for coal, and an increase in fuel costs of 1 cent per million BTU increased operating expenses by approximately $3.25 million.

As for the cost of capital, Duke Power's interest expense in 1969 was some 30 percent higher than in the previous year. Early in 1970 the company issued $75 million of bonds which carried an all-time high interest rate (up to that time) of 8.5 percent.[7]

What to do? Adding urgency to the situation was a basic fact of economic life for all investor-owned, regulated electric utilities: financial markets move and react with amazing rapidity while regulatory commissions, being subjected to a variety of political pressures, move much more s-l-o-w-l-y. Alarmed by the startling jump in the price of coal, Duke Power in July 1969, asked the North Carolina Utilities Commission for a fuel-adjustment clause that would allow the company to promptly pass along increases in coal costs in monthly bills to customers. That way the rate increases would come in relatively painless increments rather than in one jarring painful jump. Eight long months later, the Commission denied the request. During those months the cost of fuel rose even more, and the company's financial condition began to suffer. While subsequent and totally unanticipated developments (such as the Arab oil embargo of late 1973) would exacerbate an already bad economic situation, the fact was that Duke Power's inability to gain prompt, adequate relief from the regulatory commissions in North and South Carolina in 1969–1971 laid the basis for what would become a six-year ordeal for the company. "Regulatory lag" was by no means the only cause of the company's financial trouble, but it was certainly a major cause.

As the company's quarterly reports began to show a drop in earnings per share, the sharp-eyed analysts of electric utility securities—the specialists in New York and elsewhere whose daily job it was to monitor the financial health of the nation's investor-owned electric utilities—reported the bad news. At the beginning of 1970, the credit rating of Duke's first mortgage bonds (the most economical source of long-term capital) was the highest—AAA. That rating first declined to AA and then subsequently to A. The annual interest demanded by purchasers of utility first mortgage bonds rated A is from ½ percent to 1½ percent more than the interest rate on AAA bonds, depending on market conditions. As one Duke Power officer noted, "1½ percent of one hundred million dollars for 30 years is a lot of money!"[8]

What the public might not have grasped—although the members of the regulatory commissions certainly should have—was that consumers of electricity ultimately had to pay for the rising cost of both fuel and capital—if the power company was to stay in business. In other words, for Duke Power to have and keep a AAA bond rating was to the economic advantage, not only of the company, but also of the customers of the company, even if they might not fully understand what bonds and bond ratings actually were.

As Richard Pierce, a long-time officer in Duke Power's public relations department, explained, customers of the company had two vital stakes in their electric service, one short-range and one long-range. The short-range stake was to have the lowest possible electric bill each month. The long-range stake was to have reliable electric service which would prevent brownouts, blackouts, and power interruption—not just at home but at the workplace. Certain consumer groups, angry about the rate hikes of the early 1970s (especially after the rate cuts of the 1960s), focused on the short-range stake and ignored the long-range one, saying blithely, "Oh, brownouts or blackouts can't happen here!" But, as Pierce noted, they were beginning to happen in some parts of the country where electric utilities had been unable to attract new capital because they could not raise rates sufficiently to cover costs.[9]

In April 1970, Duke Power applied for an across-the-board general rate increase, this being only the second time in the company's history that it had made such a request. The company, insisting that its request was not "padded," asked for a 18 percent hike. Nearly a year later, Duke Power received permission for a 10.38 percent increase in rates. And so it would go as runaway inflation continued to plague the economy.

Meantime, electricity usage in the Piedmont Carolinas kept increasing. Not only did the Federal Power Commission predict continued, steep escalation in power consumption, but also the company's own experts and various others across the nation did likewise. Duke Power had a legal obligation to try to stay ahead of the demand for electricity, so regardless of financial problems, it believed that it had no choice but to keep adding new generating units and plants. The seventeen new units added to already existing plants between April 1967, and June 1970, were mentioned in the previous chapter, as were the Keowee Hydro Station, the Oconee Nuclear Station (with its first unit beginning commercial operation in July 1973), and the pumped-storage facility at Jocassee Hydro Station.

In 1969 Duke Power announced plans for two new, quite large generating plants: Belews Creek Steam Station near Winston-Salem, North Carolina, and McGuire Nuclear Station on Lake Norman. The Belews Creek plant, located some twelve miles northeast of Winston-Salem, would contain two generating units which would be among the largest ever built. Each generator would add 1,444,000 kilowatts to the system, so the completed plant would produce each year more electricity than the entire Duke Power system used as recently as 1965. The plant, like the other large Duke steam plants, was designed for continuous operation, that is, to supply base load.

The company chose the Belews Creek site because of economic and population growth in the northern end of the service area, from the Triad (Greensboro, Winston-Salem, and High Point) to the Research Triangle near Durham. The location in the northern end reduced investment in transmission lines, and relative proximity to the coal fields would reduce freight charges. Although the plant was estimated in 1969 to cost $289 million, because of inflation it ended up costing over $336 million by the time it became operational in 1974.

With environmental awareness and concern mounting in the 1960s and 1970s, how did people in Winston-Salem feel about having such a large coal-burning steam plant as a close neighbor? Many probably shrugged and went about their business, but the *Winston-Salem Journal* published a carefully balanced editorial that illumined a large problem. The paper saw the Belews Creek plant as presenting a classic dilemma in the nation's fight to control pollution. "That dilemma, simply stated," the paper declared, "is that present day technology will not permit us to keep buying more air conditioners, more cars, and more material goods of all kinds without pumping more filth into our air and water."

The editorial then explained, quite accurately, that Duke Power, like other utilities, was under irresistible pressure to generate more electricity. On July 21, 1969, a sweltering day when fans and air conditioners were running full blast all over the Piedmont, Duke Power's all-time record of peak demand for electricity was broken. If demand had overrun capacity and neighboring utilities had been unable to sell surplus power to Duke, power would have gone out in scattered areas of the Piedmont.

Six months later, the peak-demand record was broken again, when heaters and heat-pumps were battling to warm hundreds of thousands of Piedmont homes on a freezing day in January. And people had not stopped buying air-conditioners, refrigerators, and other power-guzzling gadgets. "But people are also up in arms," the *Journal* continued, "about industries that damage the environment; and power plants, both conventional and nuclear, are natural objects of wrath."

Like other industries, Duke Power was "throwing technology into the breach in an effort to have the best of all possible worlds—warm in the winter, cool in the summer, but freer of pollution than it is now." The trouble was "that man can't have it that way—at least not yet, and the Belews

Creek plant is a good case in point." The plant would burn an estimated 34 million pounds of coal a day. In an effort to eliminate flyash that would be pumped into the air, Duke Power planned to spend considerable money to install one of the latest anti-pollution devices, "precipitators" guaranteed by their makers to be 99.4 percent effective. Even if they performed as advertised, the *Journal* argued, that would still permit approximately 16 *tons* of fly ash to escape into the air each day.

"In a real sense," the editorial somewhat gloomily concluded, "the dilemma is not Duke's... [but] ours, and it is time we began deciding whether we want to push material progress up to the point where it requires a face mask for its enjoyment."[10]

The Winston-Salem paper struck a pessimistic note, but the *Greensboro Daily News* wrote an editorial about thermal discharge from the Belews Creek plant that was factually inaccurate. The Greensboro paper mistakenly stated that water condensed from steam would be discharged into the Belews Lake that Duke was constructing. The fact was that the water condensed from steam was not discharged but used over and over again in a closed cycle. A technically informed reader caught the error and sent an amusingly barbed letter to the editor: "... This is a good example of misleading the public by false information. You seem to want the federal government to control everybody but you, and believe me when it comes to pollution control your paper could stand a little."[11]

One of the attractive features of nuclear energy, of course, was that there was no air pollution involved. When the Belews Creek plant was first conceived, there was a debate as to whether it should be nuclear or coal fired, and there were good arguments made for both. "However, considering the importance of bringing the generation into service on schedule," Bill Lee explained, "and in view of the higher risk of regulatory delays associated with nuclear, we have decided to use coal in this instance." Lee was quick to add that "this in no way alters our conviction that nuclear power will be playing a major role in our future plans."[12]

That Bill Lee, a stalwart believer in nuclear energy, was not just whistling Dixie when he made the above statement became clear a few months later: in November, 1969, Duke Power announced its plans to build its second nuclear plant, McGuire Nuclear Station on Lake Norman. To have two nuclear-fueled generating units, the plant, when announced, was expected to be in operation by early 1977. Actually, it would not be until late 1981 before the McGuire plant went into commercial operation, more than ten years after construction had begun in 1971. It was this type of delay, caused largely by changing and ever-increasing regulations, that added so much to the costs of nuclear plants and would ultimately help bring the construction of new ones to a halt in the 1980s.

For McGuire Station's unit one, Westinghouse designed and manufactured its largest-ever generator, a record 1,305,000 kilowatts. It was not the size of the plant, however, that led a group of homeowners on Lake Norman to protect its construction, but rather it was the fact that it would be nuclear. Duke Power nevertheless stuck by its guns: "... large nuclear power plants definitely can be compatible with the environment, and the absence of conventional air pollutants is a significant advantage for nuclear power."[13]

While more nuclear power was clearly in Duke Power's future, coal-fired generation continued to be primary and absolutely indispensable. In August, 1970, however, the company's coal supply was down to 23 days, while a 70-day supply was regarded as a safe minimum. Alarmed by this situation, Duke Power decided on a course of action that it would later grow to regret: it purchased two coal mines in eastern Kentucky's Harlan County, also known as "Bloody Harlan" because of chronic and sensational labor violence in the 1930s. Lacking any experience in the management of such properties, Duke Power organized a wholly-owned subsidiary, Eastover Mining Company, to operate the two mines along with a couple of others that Duke subsequently purchased. Even with the best of results, which were not to happen, Duke Power hoped to obtain only about a quarter of its coal supplies from the Eastover mines. Despite that, however, the company hoped the new properties would help stabilize its critical supply of coal.

From the relatively halcyon days of the early and mid-1960s to the more problem-filled era that began in 1969, Bill McGuire led a fast-growing company that both encouraged the Piedmont Carolinas' appetite for electric power and worked hard to stay ahead of that demand. In April 1971, McGuire retired, and Carl Horn, Jr., became Duke Power's seventh president at age forty-nine.

From high school in Salisbury, North Carolina, Horn went to Duke University, from which he graduated in 1942. After four years in the U. S. Army during World War II, when he went from the rank of private to that of captain, he returned to Duke University for a law degree; after practicing law in Charlotte for a few years, he joined Duke Power in 1954. Moving steadily up the corporate ladder, Horn became executive vice president, while continuing as general counsel, in 1970.

When Horn became president in 1971, Duke Power's problems, while already serious, were soon to become much worse. All the forecasts, however, were for an accelerating demand for electricity, and in July 1972, the company announced plans for its third nuclear plant.

The Catawba Nuclear Station would be located on a peninsula in Lake Wylie, about three miles north of the re-built dam where the company had started in 1904. Identical in size and design to the McGuire plant, Catawba Nuclear Station was originally estimated to cost $700 million (but ended up costing more) and represented the largest single industrial facility in the history of South Carolina. Construction began in May 1974, and more than eleven years later in June 1985, Catawba's unit 1 (1,145,000 kilowatts) began commercial operation. Although the final cost was more than had been estimated, Duke Power's "do-it-yourself" construction policy again paid off: it was the lowest cost per kilowatt of any initial unit of a nuclear plant of similar vintage in the United States. (Unit 2 began commercial operation in August 1986.)[14]

Double-digit inflation, triggered partly by Federal spending and an increasing deficit in the Federal government's budget, was the underlying cause of the electric utilities' mounting problems. The more immediate cause, however, was rooted in "regulatory lag." That is, by the time Duke Power and most other utilities gained permission to increase rates—and the increases were usually considerably less than requested—rising costs had already outpaced the granted increases. Between July 1969, and early 1972, Duke Power filed and participated in hearings on six applications for rate increases before three regulatory commissions (the two Carolina state commissions and the Federal Power Commission, which had jurisdiction over wholesale rates). In order to continue the massive construction program that was underway, Duke Power had to file for another rate increase in 1972.

Trying to explain to increasingly unhappy customers why electric rates had to increase, Duke Power ran a series of full-page advertisements in the larger newspapers of the Carolinas that focused separately on each of these messages:

- "The interest on our construction loans increased $13,000,000 in one year."
- "The fuel we used in 1970 cost $49,000,000 more than the fuel we used in 1969."
- "The coal fired power plant now being built [at Belews Creek] will cost 37% more than our last one [at Marshall Steam Station]."
- "We're spending $89,000,000 to protect the environment."
- "Today you're using 10 times as much electricity as you did 30 years ago. But you're only paying 58% as much per kilowatt hour."

Carl Horn both apologized and tried to explain: "We especially regret the prospect of higher rates, after 20 years of achieving real growth with rate reductions." Increases were essential, however, for the necessary construction program.[15]

Duke Power did not, of course, just sit, wring its hands, and complain about cruel circumstances beyond its control. For one thing, the company moved to increase its revenues as much as possible from non-regulated, non-utility-related subsidiaries. While the larger pay-off would come in future years, in 1969 Duke Power transferred title to some 300,000 acres of its non-utility lands to its wholly-owned subsidiary, Crescent Land and Timber Corporation.

One of Crescent's early attempts to add to Duke Power's bottom line did not turn out well. Crescent was a participating partner in the development of a large amusement and resort complex near Charlotte known as Carowinds. Unfortunately for Duke Power, that venture did not pan out, and the company lost money on it. Other ventures ultimately proved more profitable as Crescent, with its own financing obtained independently of Duke Power, sold or leased land for commercial development and participated in a number of such developments through joint ventures and part ownership. In the long run, up-scale developments on Lakes Keowee and Norman would prove especially rewarding for Crescent—and Duke Power.

The "long run," however, could not help with the immediate financial crunch of the early 1970s. Because a notably decreasing share of construction costs could be covered by retained earnings, Duke Power increasingly had to look to the sale of stocks and bonds to obtain needed capital. To encourage those sales, especially after the Duke Endowment became barred after 1969 from helping to provide capital for Duke Power, the company acted to increase the number and broaden the base of its shareholders.

Duke Power promoted William H. Grigg to the position of vice-president for finance in 1970. Born in Shelby, North Carolina, in 1932, he grew up in Albemarle, North Carolina, and graduated from Duke University in 1954. After two years in the United States Marine Corps, he returned to Duke University for a law degree, and practiced law in Charlotte for a few years before taking a job with Duke Power in 1963. Destined to be the last president and chief executive officer of Duke Power before it changed its corporate identity in 1997, Grigg also played a prominent role alongside Horn and others in coping with the daunting problems of the early 1970s.

Prior to offering 2½ million shares of Duke Power common stock in 1970, Grigg and a few other company representatives travelled across the country to familiarize potential investors with the company. As a result of those endeavors, the company boasted in early 1971 that it had become a "truly national corporation" with shareholders in every state except Alaska and more than 2,000 shares held by foreign investors. As would be expected, the company's two home states topped the shareholders distribution list, with almost 2 million shares held in North Carolina and over ¼ million in South Carolina. But following the financial team's visit to California, the number of Duke Power shareholders there jumped from 121 to 718, and there were similar increases in Illinois, Massachusetts, and elsewhere. Too, 8,511 employees—81 percent of those eligible—were participating in the employee stock purchase-saving program.[16]

Sadly for Duke Power shareholders, however, the news in the early 1970s was not good. Just as the company's falling bond rating, which was discussed earlier, reflected and aggravated the company's financial problems, so did the precipitous drop in the price of Duke Power common stock on the New York Stock Exchange. As 1970 ended, the closing price on the Exchange of Duke Power Common Stock was $24.75 per share, or 15.7 times the 1970 earnings of $1.57 per share and 140% of 1970 book value per share of $17.64. (When a stockholder calculates the amount of equity behind each share, the result is called "book value." It is determined by dividing the amount of stockholders' equity by the number of shares outstanding.)

By the end of September 1974, however, Duke Power common closed at $10.50 per share, down $14.25 from its 1970 price. The $10.50 price was only 5½ times the earnings as of September 1974, of $1.89 per share, and only 51 percent of the stock's book value of $20.41 per share. (By early 1975 the price rose to $14.00 a share.)[17]

What this meant, of course, was that holders of Duke Power common stock saw its price plummet drastically. Moreover, as the company was forced to raise capital by selling more common stock at well below book value, that further diluted the value of the stock.

One should certainly understand that Duke Power, for all its financial problems, was not alone. By September 1974, prices for electric utility stocks in general had fallen by 36 percent. Many companies were in much worse condition than Duke Power. Consolidated Edison in New York, for example, in April 1974, did the unthinkable for an electric utility: it failed to pay its common stock dividend. According to an utility analyst and historian, Con Edison's action "hit the industry with the impact of a wrecking ball." It "smashed the keystone of faith for investment in utilities: that the dividend is safe and will be paid."[18]

While Duke Power, unlike Consolidated Edison, kept paying the annual dividend on its common stock, the dividend fell to $1.40 in 1970 and remained stuck there for five years. Moreover, in order to pay even that amount, the company finally had to borrow money.

Because of an unforeseen development in the Middle East, economic conditions in the United States took a dramatic turn for the worse in the fall of 1973. Angered by the United States' support for Israel in the Arab-Israeli conflict at that time, Saudi Arabia and its allies declared an embargo on oil shipments to the United States. Not only did petroleum prices soar upwards, but also acute shortages of gasoline soon had automobile-loving Americans frustrated as they had never before been. Motorists were forced to form long lines of their cars at service stations—if any gasoline should be available—and then could only purchase a limited amount at an unprecedentedly high price. Thoroughly spoiled by a long history of abundant, cheap energy, Americans for the first time got a bitter taste of what people in most other parts of the world had long experienced: scarce and expensive energy.

Duke Power was fortunate in that it made little use of oil-fired generation. In the Northeast, on the other hand, some 60 percent of the generating plants were oil-fired, and the dramatic jump in oil prices following the embargo played special havoc there. The oil embargo, coming on top of the inflation that had begun to ratchet upwards from 1969 on, threw the whole economy into the 1974–75 recession, a downtown that many observers believed to be the worst since the 1930s.

Perhaps some perspective may be gained on Duke Power's problems by the later comments of a couple of business leaders about the recession of 1974–75. Hugh McColl of Charlotte, one of the nation's top banking leaders in the late twentieth century, later declared, "During the 1974–75 recession, NCNB [North Carolina National Bank] nearly became insolvent." The bank had bet wrong on interest rates, McColl explained, and got caught with too many bad loans on its books. "That period separated the lions from the lambs," he concluded, and "it sobered us on many fronts."[19]

Likewise, Robert Rubin, Secretary of the Treasury in President Bill Clinton's administration and formerly with Goldman, Sachs and Company, had this to say concerning the recession and stock market decline of 1974–75: "The difference between the people who were smart and shrewd, and the people who weren't really smart and shrewd?" The former "lost a lot of money," and the latter "got wiped out." Success, he concluded, "meant that you managed extremely difficult forces effectively enough to . . . come out at the other end."[20]

According to Rubin's definition, Duke Power ultimately proved to be a "success," for it did "come out at the other end"—but it was strictly a touch-and-go situation by 1974–75. The managers of electric utilities across the country, including Duke Power, as well as the state regulatory agencies and the Federal Power Commission were still predicting a sharply rising rate of electricity usage through the rest of the century. While the oil embargo had introduced a totally unexpected twist to the energy situation, that the embargo would gradually help plunge the nation into a severe recession was by no means clear as 1974 began.

Those facts should be kept in mind as background for a fateful decision that Duke Power made early in 1974. About the best that can be said about the decision is that Duke Power, like so many of the nation's other electric utilities, failed to realize in time that a whole new era in electricity usage had begun. President Richard Nixon first called for an all-out national effort to conserve energy following the Arab oil embargo. Then, following his resignation in 1974, President Gerald Ford reiterated the same urgent message. The American people, responding partly through patriotism and partly through frugality as the price of fossil fuel and electricity soared, began to do something they had never before attempted on a massive, nationwide scale—conserve energy.

Duke Power's *Annual Report* for 1973, which was written early in 1974, reflected much of the dominant thinking of the confused time. The report emphasized, for example, that the company was actively engaged in a variety of energy-conserving procedures, such as computer programs to analyze distribution circuits in order to reduce distribution-system losses. Other computer programs aimed at permitting the utilization of the most efficient plants for base load, minimizing (and in some cases eliminating entirely) the use of generating equipment that required fuel oil or natural gas.

The company was designing residential rate schedules to encourage greater home insulation and industrial rates to encourage limitations on peak-period electrical demands. Duke Power had also undertaken an extensive program to educate customers in the efficient use of electricity. The program included the promotion of "heat recovery" energy systems for commercial and industrial uses, systems which redistributed heat created by lighting, machines, and people, thus providing the ultimate in energy conservation.

The company had also initiated informational and educational programs with builders in the residential field and with consulting architects and engineers in the commercial and industrial areas. Duke's Home Service Department was giving demonstration programs on the efficient and economical uses of electricity, and direct-mail and mass-media advertising disseminated information designed to help customers reduce wasteful use of electricity in the home. "While efforts are continuing to encourage prudence in the end use of electricity," the report concluded, "we realize that substantial savings of fuels can best be realized through improved efficiencies in the generation, transmission, and distribution of power."[21]

While Duke Power had clearly begun to push conservation by late 1973, it also remained strongly committed to nuclear power, perhaps even more so after the oil embargo. The 1973 *Annual Report* spelled out the rationale for nuclear and noted that in the early 1960s Duke had set as one of its chief goals the development of a generating system that would rely heavily on nuclear power. Since that time the company had invested more than $700 million in nuclear generating facilities, and a total of $6.4 billion more was committed to that goal in the eleven years through 1984.

Although predicated on economic and environmental considerations, the report explained, the decision in favor of nuclear power had an impact on the conservation of fossil fuels that was not appreciated until after the oil embargo and price hike, when concern over world energy resources reached crisis proportions. In addition to conserving fossil fuels, moreover, nuclear power plants had other advantages over conventional coal-fired generating plants. Nuclear power plants were more compatible with the environment. There were no emissions of "flyash" or products of combustion into the atmosphere. (And while sulphur dioxide was certainly a concern by 1973, no one had yet begun to talk or worry about "greenhouse gases," especially carbon dioxide, and their possible contribution to the controversial phenomenon of "global warming.") The report went on to argue that the small quantities of radioactive waste from a nuclear plant could be handled with less environmental impact than could millions of cubic feet of waste particulates that were collected annually in a modern coal-burning plant.

Nuclear plants also had a distinct economic advantage over coal-burning plants, and Duke Power believed the nuclear plants would help hold down future increases in power costs. The cost of constructing a nuclear plant on the Duke system (where construction costs were substantially below then-current industry averages) was about 40 percent greater than the construction costs of a comparably sized coal-burning plant. The initial cost of a plant, however, was only one factor in determining the cost of electricity produced by the plant, for the cost of fuel was also a large factor.

The economic advantage of nuclear over fossil fuels meant that electricity produced by a nuclear station such as Oconee Unit 1, which began commercial operation in July 1973, cost about 25 percent less than that produced by a coal-burning plant. "Our customers will have the benefit of a cleaner environment as well as lower rates than would have been the case if all generation were from fossil plants," the report concluded.[22]

These considerations then—the anticipated high rate of demand for electricity, the shortage and high price of oil and gas, and the environmental and economic advantages of nuclear over coal—lay behind Duke Power's decision in 1973–74 to plunge even further into the building of nuclear plants. Oconee was already built; construction on McGuire began in 1971; and construction on Catawba would begin in May 1974. Now, in February 1974, Duke Power proudly and boldly (and, as later events would reveal, unwisely) announced plans for an unprecedented "Six-Pack": three generating units at the Thomas L. Perkins Nuclear Station to be built on the Yadkin River near Mocksville, North Carolina, to cost an estimated $1.2 billion; and three more units at the Cherokee Nuclear Station to be constructed on the Broad River near Blacksburg and Gaffney, South Carolina, to cost an estimated $3.5 billion.[23]

At the same time that Duke Power made its large commitment to additional nuclear plants, it also announced in February 1974, that it would build an additional pumped-storage hydroelectric plant on Lake Jocassee in the Keowee-Toxaway development. Construction on Bad Creek Hydro Station would be delayed, however, and not begin until 1981. For the company's twenty-seventh hydro plant, Duke Power engineers and construction workers again pulled off a most imaginative feat. While the much-touted nuclear "Six-Pack" would never actually materialize, Bad Creek Hydro Station would eventually play an important role in helping Duke deal with the critical matter of peak load in the 1990s. Since Bad Creek's construction would not began until 1981, however, the discussion of the project will be held for a later chapter.

By 1974 and its severe economic recession, Duke Power's difficult financial situation was being seriously complicated by trouble from another quarter—the company's coal mines in Harlan County, Kentucky, which it had purchased in 1970. Lacking any expertise in coal-mining, Duke Power employed an experienced mining engineer, Norman Yarborough, to head its new wholly-owned subsidiary corporation in the coal fields, Eastover Mining Company and Eastover Land Company. In 1971, Duke Power purchased a third mine in adjoining Bell County, Kentucky, and the hope was that when the three mines were fully developed—and they were not in good shape when Duke bought them—they would furnish about 25 percent of the company's annual coal requirements.

The first two mines that Duke Power had purchased, Brookside and High Splint, had contracts with the Southern Labor Union, and Eastover (the Duke subsidiary) took over those contracts. The top rate of pay had been $18 per day, but when Eastover began operations the top rate was $28 per day; by 1974 the top rate had risen to $48 per day. That, as Carl Horn explained, was "more money than the average Charlotte school teacher, policeman or top Duke Power servicemen or lineman earned in 1973."[24]

When the contract with the Southern Labor Union expired at the Brookside mine in mid-1973, the miners voted, in a hotly contested election, to be represented by the United Mine Workers of America (the UMW). (The contracts with the Southern Labor Union at the other two mines would not expire until mid-1975.)

The UMW, once led by the redoubtable John L. Lewis, had been powerful in eastern Kentucky from the late 1930s into the 1950s. As coal prices in the 1950s slumped as low as $4.50 a ton, however, scores of mines closed when they were unable to meet the union pay scale of $26 a day. By the mid-1960s the once-powerful UMW had lost its grip in eastern Kentucky and had been ousted from all but one or two of the region's mines. Then when coal prices began to soar from 1969 onward, the $32-a-ton coal brought a promise of new prosperity to the Kentucky coal fields.[25]

An additional twist came from the fact that in 1973 the UMW, once plagued by scandal and violence, had a new leader pledged to reform, Arnold Miller. (His predecessor as UMW president, W. A. (Tony) Boyle, was in prison for the murder of his rival in a previous UMW presidential election.) Determined to reestablish the UMW in eastern Kentucky, Miller and his associates were prepared to spend a vast amount of money and to play a type of hard ball that would prove to be something hitherto unknown by Duke Power. From 1938 on Duke Power had contracts with the International Brotherhood of Electrical Workers, and there had been no strikes.[26] Yet Duke Power was about to be unfairly assailed in some quarters as anti-union.

From the beginning, Duke Power officers in Charlotte left negotiations with the UMW completely in the hands of Eastover's president, Norman Yarborough, and his local attorney, Logan Patterson. With the benefit of hindsight, one might argue that such an arrangement was short-sighted, for Duke Power would prove to have certain larger interests involved that transcended the matter of a labor contract at a relatively small coal mine in Kentucky. Yet Yarborough and his attorney, Patterson, understandably perhaps, focused exclusively on the immediate situation in Kentucky.

At any rate, Eastover and the UMW began bargaining sessions about a new contract in late June 1973. When the negotiations failed, the majority of the Brookside miners went on strike on July 30 to put pressure on Eastover—and Duke Power. Bargaining sessions, some fourteen in all, continued until late November 1973, when each side filed charges with the National Labor Relations Board accusing the other of bargaining in bad faith.[27]

Since Eastover's wage rates were higher than those of the UMW, money was not an issue in the negotiations. They stalemated because from the outset the UMW insisted that Eastover would have to sign the same three-year contract that the UMW had made with the Bituminous Coal Operators Association in 1971. Yarborough and Patterson balked at these features of that contract: (1) it would cover all operations of the employer on any properties where the employer mined coal; (2) it lacked a no-strike clause; (3) promotions would be based on seniority regardless of ability; (4) it lacked a management-rights clause; and (5) it gave the UMW's safety committee the unilateral authority to shut down the employer's mine.[28] Subsequent events would reveal that most of the above matters were more negotiable than they first appeared to be, but Eastover's insistence on a no-strike clause was the crucial sticking point.

In Harlan County, where memories of bitter and bloody labor battles in an earlier era ran deep, small-scale violence began almost as soon as the strike began. One of the reasons was that the coal miners and their families, as well as the community at large, were deeply divided about the matter, some being just as hostile to the UMW as its partisans were hostile to the Southern Labor Union. Under a court injunction against "mass and violent picketing," some of the anti-UMW miners employed at Brookside attempted to return to work in the mine under the protection of a court order. Although the non-striking miners were escorted by officers of the Kentucky State Highway Patrol, miners' wives and other sympathetic women lay down in the road to block the patrol cars while other women vocally assailed the patrolman as well as attacked them with "switches" and sticks when they tried to remove prostrate women from the roadways. (By 1973, of course, the tactics of mass civil—and sometimes not-so-civil—disobedience had been demonstrated on the nation's television sets for a good decade or more.) Although the sheriff served many of these women with citations and some were convicted of defying the injunction against violent picketing, the angry demonstrations did not cease but eventually spread to other Eastover-owned mines.[29]

Fearing that continued violence by these picket-line "peacekeepers" would lead to serious bloodshed, Duke Power ordered Yarborough to close the Brookside mine except for maintenance and security personnel. (Without regular maintenance, such as pumping operations, the mine would soon become inoperable.)

The real battle about the Brookside strike, however, was not to be fought in Kentucky but in the Carolinas and on Wall Street. The UMW, capitalizing on Duke Power's relationship with the Duke Endowment, waged a masterful, even if ethically dubious, public relations war against Duke Power, and in the process considerably embarrassed both the company and, to a lesser extent perhaps, the Endowment.

Shifting away from the actual issues that divided the two contestants, the UMW launched an all-out "humanitarian" campaign that essentially blamed Duke Power for not having solved historic socioeconomic problems in Harlan County that had existed for decades. (And, in fact, the UMW had long been a major force in Eastern Kentucky whereas Duke Power arrived on the scene in 1970.) Nevertheless, in a letter to all of the trustees of the Endowment, Arnold Miller explained that he was writing because the Endowment was the major and controlling stockholder in Duke Power. "So I know you will be disappointed to learn that in Harlan County, Kentucky," the UMW leader declared, "Duke Power Company has shown a callousness to human needs and suffering that flies in the face of everything the Duke Endowment stands for." While the Endowment was concerned about child care, the letter continued, in Harlan County Duke Power housed the children of its coal miner employees in "ramshackle coal camp houses with no indoor toilet, running water, or central heating."[30]

The long letter (five single-spaced pages) went on in this vein, but to take only the matter of housing, the truth was a bit more complicated than Miller suggested. When Duke Power acquired the first two Harlan County mines, there were a large number of ramshackle, company-owned houses. Duke Power, or rather its subsidiary, Eastover, demolished a hundred of those rundown houses at Brookside mine and another hundred at the other two mines. The company then spent several thousand dollars for roofing, screening and paint on the remaining forty houses at Brookside, and the improvement was sufficiently impressive that Harlan County gave Eastover its annual award for beautification. It was also true, however, that of the forty remaining houses at Brookside, only fifteen had indoor plumbing. The others were scheduled for demolition, to be replaced by a mobile-home park with sewage facilities. In other words, Eastover had acted and was continuing to act to rectify an admittedly deficient housing situation, one that had long existed in a generally depressed region. (Housing was in short supply in Harlan County , and of approximately 12,000 units, about 40 percent lacked some plumbing facilities.)[31]

"I was horrified to hear of the appalling conditions at Brookside camp," Doris Duke, who had been a trustee of the Duke Endowment since 1933, wrote the president of the UMW. She noted that she was only one of the fifteen trustees of the Endowment and had "very little influence," but she had telephoned her fellow trustee and cousin, Mary Semans, who had also received the UMW letter. "She too was extremely distressed to learn of these conditions," Doris Duke continued, "and said they planned to have one of the Board Members of the Endowment meet with you for a personal inspection tour." Doris Duke concluded by stating that she had been assured that she would be kept informed as to what progress was made "to alleviate this situation."[32]

Doris Duke claimed to have "very little influence," but, for many reasons, that was certainly incorrect. Carl Horn soon wrote to assure her that he would be happy to meet with her at a time and place of her choice to discuss the labor situation at Duke Power's mines in Harlan County. He believed that some recent developments suggested the possibility of a settlement, for he had finally met with Arnold Miller and hoped the negotiations would soon resume. Even if a settlement were reached, Horn thought Doris Duke would be interested in the company's long-range plans for improving the miners' living conditions. Duke Power had bought into "an economically depressed area," Horn noted, and while it had "improved the economy of the county, much remains to be done, particularly in the areas of housing, medical care and family planning."[33]

Duke Power would have been happier if the UMW had confined itself to letters to Endowment trustees, but such was not the case. With a sizable war chest, the UMW was determined to win the contest with Duke Power as a necessary first step toward vanquishing its rival in eastern Kentucky, the Southern Labor Union. Accordingly, the UMW launched an intensive (and Horn said "scurrilous") attack on Duke Power, not only in the Carolinas, but also in the nation at large, and especially on Wall Street.

The timing here was crucial, for in late 1973 and early 1974 Duke Power was fighting for an absolutely essential rate hike in both Carolinas. The company's financial condition had steadily deteriorated from 1969 onward, and without the requested 17 percent rate hike, Duke Power, in the words of Bill Grigg, then the company's general counsel, faced a "disastrous financial condition."[34]

As an additional way of putting pressure on Duke Power, the UMW placed in all the major newspapers of the Carolinas full-page advertisements opposing the utility's request for the rate hike and charging that its refusal to sign a contract with the UMW was somehow responsible for the company's financial dilemma. By the spring of 1974 the UMW had placed a total of ninety-one full-page advertisements in sixteen Carolina newspapers, plus a number of smaller advertisements. The UMW had spent over $30,000 for space in the *Charlotte Observer* alone.[35]

Opening a second front, the UMW placed advertisements in the *Wall Street Journal*, the *New York Times*, and financial publications such as *Bar-*

ron's with the purpose of discouraging potential investors in Duke Power securities. Some newspapers allowed the UMW to hide behind so-called consumer groups such as "Carolina Action" and "Palmetto Action," but other papers required that the advertisements carry the statement that they were paid for by the UMW. As an example, one of the UMW advertisements bore the heading "Guess Who's Paying for the Brookside Strike?" and went on to suggest that Duke Power customers bore the costs and should "Fight Duke Power's Rate Increase—Back the Brookside Miners." The advertisement charged, among other things, that customers were "Paying as many as 50 strikebreakers $48 a day a piece—not to mine coal—but just to cross the Brookside picket line." In a booklet prepared for Duke Power employees and any interested others, the company answered the charge: "Not true. Eastover has never employed so-called 'strikebreakers'. A number of Brookside employees asked to continue at their jobs and were permitted to do so under court order—until it became obvious that continued violence by the picket line mobs would lead to bloodshed. To prevent this, the mine was closed down except for supervisory, maintenance and security personnel."[36]

Another UMW advertisement heralded the news that "Members of Congress Tell Duke Power 'Settle the Brookside Strike!" Duke Power pointed out the interesting fact that of the forty-one congressmen signing the advertisement, not a single one was from Kentucky where the Brookside mine was located or from the Carolinas, the states served by Duke Power. Eleven of the forty-one signers, in fact, represented California, "a state not noted for its coal mines."[37]

Enthusiastically assisting the UMW in its campaign were groups of student activists at Duke University and the University of North Carolina at Chapel Hill. With the war in Vietnam eliminated as a focus for student activism, these groups, backed partly by UMW funds, worked through the Public Interest Research Group (PIRG), a nationwide organization inspired by Ralph Nader. Placard-carrying students assailed Duke Power in picket lines and protested against rate hikes at various strategic locations throughout the Carolinas.

While militant students and others protested in the Carolinas, Arnold Miller and the UMW took the fight to Wall Street. In a letter to the head of New York's First National City Bank, Arnold Miller began by stating that he wanted to express his concern about the bank's holdings of stock in Duke Power. After giving the UMW version of the Brookside strike, Miller went on to explain that, aside from the social implications of being a large stockholder in Duke Power, the bank should know that there was widespread citizen opposition in both Carolinas to Duke Power's requested rate increases. Nine or more different organizations in North Carolina and ten in South Carolina ("Carolina Action," "Palmetto Action," Duke University PIRG, etc.) were reported to have formally intervened to oppose the rate increases, and the UMW had retained a public utility consulting firm in Washington to provide expert testimony on behalf of the intervenors.

The UMW's nationwide "informational" campaign against Duke Power securities, Miller explained, had brought results: fifty-two national labor organizations and four national church denominations had pledged to dispose of holdings in Duke Power and not to buy more until the Brookside strike was settled. The sixth largest stockholders in Duke Power, the Ohio Public Employee Retirement System, had informed the company that it, too, would purchase no more Duke Power securities until the strike was settled. "I am sure that the American labor movement would welcome the news—as the United Mine Workers would," Miller concluded, "that your company intends to reevaluate its position as a major shareholder in Duke Power."[38]

Even before writing the above letter, Miller and the UMW had struck a telling but legally dubious blow at Duke Power. In March 1974 the company filed a registration statement with the Federal Securities and Exchange Commission concerning five million additional shares of its common stock to be offered for sale the following month. On April 16, 1974, as Carl Horn was about to go to New York to negotiate the terms of the stock offering with the underwriters, he received a telegram from Arnold Miller saying, in effect, that if Duke Power did not immediately (i.e., within the next twenty-four hours) agree to settle the strike at Brookside, then the UMW felt it had the "responsibility to inform potential investors and the general public about the long-term implications of the strike, our participation in the upcoming hearing" on Duke Power's request for another rate increase, and the company's financial condition. "We are taking the necessary steps to proceed in the public forum," Miller's telegram concluded, "steps which we will feel forced to take in the absence of an immediate good faith response by you to this telegram."[39]

When Horn and other Duke Power officials met with the New York underwriters, the latter recommended that the offering be reduced to four and a half million shares, but all agreed that the sale should proceed despite Miller's telegram. On April 17, 1974, the UMW took a full-page advertisement in the *Wall Street Journal* attacking Duke Power not only about its position in the Brookside affair, but also about its allegedly unsound financial condition. Unfortunately for Duke Power, the UMW advertisement used out-dated data and failed to offer information that considerably brightened the picture for the company.

The crux of the matter was that early in 1974 the regulatory commissions in both the Carolinas, despite the loud protests of the PIRG students and others, had granted the crucially important rate increases. If the comparative statistics used in the UMW's advertisements had reflected those rate increases, Duke Power's financial condition would have placed it much nearer the top of the list of 110 utility companies. The 14 percent return on common equity made possible by the rate increases would have placed Duke Power as the 20th out of 110 utilities rather than as the 107th, as shown in the UMW's advertisement.

There were many other such distortions and inaccuracies, and one of Duke Power's lawyers summarized the matter thusly: "The situation described above is a unique and novel development in the law of federal securities regulation. What has occurred is that a union, in an effort to obtain its objectives in a labor dispute, published false and misleading statements about an issuer [of stocks] at a time when the issuer was offering a very substantial amount of its securities to the public and [was] thus severely limited in its ability to respond to such allegations." If this action of the UMW should not be challenged, according to counsel for Duke Power and the Duke Endowment, then a major objective of federal securities law—the sale of securities to the public in an orderly fashion upon the basis of full and fair disclosure—would be defeated.[40] Although Duke Power, through its lawyers, did protest to the Securities and Exchange Commission, nothing could be done about the after-the-fact situation. The company's stock offering of April 17, 1974, was successfully issued despite the UMW's advertisement.

Although Duke Power won some early victories in the Kentucky courts and even from the National Labor Relations Board (NLRB), in May, 1974, an administrative law judge of the NLRB ruled against Eastover and Duke Power. He concluded that "Eastover through its knowledgeable chief negotiator, labor attorney [Logan E.] Patterson, was well aware of the Union's ban for more than a quarter of a century of any no-strike or penalty provisions in its National Agreements and that Eastover's attempt to get such a no-strike penalty clause knowingly foreordained the negotiations to failure."

The judge went on to note that the UMW over its long history had found that it had no practical means of preventing spur-of-the-moment wildcat strikes at scattered coal mines throughout the nation. Moreover, the UMW believed that the acceptance of financial liability for such unauthorized strikes could well wreck the union financially. The UMW did, however, call attention to a series of decisions of the United States Supreme Court which held that "even in the absence of an express no-strike clause, a contractual commitment to submit disagreements to final and binding arbitration gives rise to an implied obligation not to strike over such disputes."[41]

Duke Power had every right to appeal the above decision—to the full NLRB and even in the federal courts—for the nation was in a serious "energy crisis," productivity was a major problem in the coal mines, and wildcat strikes contributed significantly to that problem. On the other hand, an appeal could drag on for many months, even years, and tension was rising dangerously in Harlan County. UMW miners and sympathizers were picketing Eastover's other mines (besides Brookside) and attempting to shut them down. Would Duke Power appeal the ruling?

The *Charlotte Observer*, long a friendly neighbor of Duke Power, hoped that Duke Power would not appeal and that the struggle could somehow be ended. Describing Duke Power as "one of the nation's most progressively managed power companies," the *Observer* also noted that Duke Power was the nation's third largest user of steam coal. That meant that it and its Carolina customers were deeply involved in the lives of the people in the coal-mining areas—"a wretched part of the country that is bleak with poverty, dilapidated houses, bad health and polluted streams, with people largely ignored by the country [at large] that benefits from their plight."

Noting the increasing tension and violence in Harlan County, the *Observer* feared that an appeal of the NLRB judge's finding could drag on indefinitely while both the miners and Duke Power suffered losses. "Duke's loss," the editorial argued, "has to do, in part, with its reputation." The UMW attacks, while at times "erroneous" and at times "simply tough," had taken a toll, and UMW-funded challenges to Duke Power's requests for rate increases further complicated a difficult but urgent matter. Duke Power's name was not one, the *Observer* suggested, "that should be identified in the future with the kind of 19th century managements that have been numerous in the coal-mining country." The paper hoped that the UMW, seeing the advantages of dealing with a better kind of management, would bend to reach a solution and that Duke Power would do likewise, for "an accord between the two, rather than a protracted and possibly bloody standoff, could benefit eastern Kentucky and the Carolinas as well."[42]

While Duke Power did not appeal the NLRB judge's ruling, neither did it act to secure a quick settlement. Arnold Miller and the UMW, for their part, obviously felt encouraged by the ruling in their favor, and in the summer of 1974 began to talk of a nationwide walkout of UMW miners unless the Brookside strike were settled. Such a walkout, coming in the midst of a critical energy crisis, was the last thing that the new administration of President Gerald Ford needed or wanted. And it was that strong position of the Ford administration that finally led to a settlement of the Brookside strike.

William Ussery, an official in the Department of Labor (who was later to become President Jimmy Carter's Secretary of Labor), summoned representatives of Duke Power and the UMW to a meeting in Washington on August 28, 1974. Carl Horn, Bill Grigg, and George Ferguson (an associate general counsel) represented Duke Power, and Arnold Miller and Chip Yablonski (general counsel) spoke for the UMW. (Norman Yarborough, delayed by a court appearance, joined the Duke group after midnight.) Ussery, according to Bill Grigg's later account, informed the group at the outset that "unless an agreement were reached by 9:00 a. m. the following morning we were to go to the Oval Office for a meeting with the President," who was "insistent that the strike be settled."

Throughout the afternoon and evening of August 28 and into the early hours of August 29, Ussery shuttled back and forth between the Duke Power trio and the UMW team. For a number of hours there was no progress, despite Ussery's alternating cajolery and threats—"the ultimate threat being a trip to the Oval Office for an enforced settlement 'in the National Interest.'"

Finally around 4:00 a. m., Ussery informed the Duke Power group that the UMW would agree to a contract that "embodied some deviation from the national contract," although there would not be a no-strike provision. Given what amounted to a slight face-saving concession, the Duke Power group agreed to settle. They then met face-to-face with Miller and Yablonski "and executed a contract—shortly before dawn on August 29 [1974]."[43]

A long, embarrassing ordeal for Duke Power had finally ended with what the *Washington Post* headline termed a "major victory for UMW."[44] The strike had cost Duke Power, in monetary terms only, about $300,000, while the UMW had spent $1.5 million; Kentucky lost over $1 million in taxes.[45] A couple of years after the settlement, a Duke Power officer (probably Carl Horn) wrote a memorandum in which the fact was noted that the company had been much disappointed in the production at Brookside, which was far beneath that of other Eastover mines where the labor contracts were with the Southern Labor Union. In fact, employees at two other Eastover mines (Highsplint and Arjay) subsequently had the opportunity to choose between the UMW and the Southern Labor Union, and chose the latter. Another small mine started up by Eastover in Wise, Virginia, did vote for the UMW by a slight margin.[46]

In 1982, after the North Carolina Utilities Commission prohibited Duke Power's full recovery of the cost of coal from its Eastover mines, the company prepared to take a $30 million loss by selling the Eastover mines and lands. Thus Duke Power finally got out of its ill-fated venture into coal mining.

The company learned at least one important lesson from the episode: as it strove diligently from the late 1970s onward to strengthen its financial base through various non-regulated, auxiliary ventures, it would do so only where it could employ its own expertise as well as its money. In other words, Duke Power wanted to avoid ever again becoming involved with an enterprise where only its money—and not its rich experience and abundantly talented personnel—would play a role.[47]

By mid-1975, as the worst of the national recession ended, Carl Horn happily reported that he and his associates could at least begin to see a light at the end of the gloomy financial tunnel. Conditions did finally improve on many fronts, but then in 1979 an accident at a nuclear plant in Pennsylvania, Three-Mile Island, would throw the whole future of nuclear-powered generation into question.

William B. McGuire

Carl Horn, Jr.

Douglas W. Booth

Thomas L. Perkins

Oconee Nuclear Station in the Keeowee-Toxaway complex.

Workers stand beside giant turbine at Cowans Ford Hydro Station.

Cowans Ford Hydro Station, with a small portion of Lake Norman behind the dam.

A recreation area in the Keeowee-Toxaway complex.

Boating on a Keeowee-Toxaway lake.

B. B. Parker

One of three Edison Awards won by Duke Power.

Preparing to sail on Lake Norman.

Bill Lee (left) greets Richard Pfaehler at Keeowee-Toxaway.

Poles and transformers downed by Hurricane Hugo in 1989.

Another victim of Hurrican Hugo.

Linemen coping with the consequences of an ice storm.

Bill Grigg (left) and Bill Lee.

Richard B. (Rick) Priory

CHAPTER VI

CLIMBING SLOWLY OUT OF A FINANCIAL HOLE, 1975–1985

As a wise person has observed, the First Law of Holes is this: when in one, stop digging. Because of circumstances largely beyond Duke Power's control—double-digit inflation, a nationwide economic recession, and inadequate rate levels being the foremost factors—the company found itself in a perilous financial condition in 1974–1975. Obligated by law to anticipate and then meet the electric power needs of its fast-growing service area, Duke had a massive construction program underway. Yet the cost of borrowing to provide the capital needed to pay for the construction program had risen dramatically, just as had the cost of coal. With Duke's common stock selling far below book value, the company could barely hang on to its single-A bond rating, which itself was low enough to increase significantly the cost of borrowing additional funds.

Fortunately for Duke Power, its management team responded to the crisis quickly and in a variety of imaginative, constructive ways. Climbing out of the financial hole, however, was a tricky, extended affair. First, the company announced late in 1974 that it was revising the construction schedule for the ten new nuclear units that had been planned—two at McGuire, two at Catawba, three at Cherokee, and three at Perkins. Although the electric utilities of the nation were charged by law with responsibility of providing a reliable source of power for their customers, as Carl Horn explained, "many companies now find themselves in the position of being economically deprived of the means by which to meet that responsibility." Duke Power was, therefore, one of many utilities being forced in 1974–1975 to make significant cutbacks or slow-downs in expansion plans. "After thoroughly investigating all available means of financing," Horn continued, "we concluded that critical conditions existing in the financial markets made it impossible to maintain the former construction schedule."[1]

The revised construction schedule, delaying the projected completion dates of the various generating units (but not yet canceling any of them), would reduce Duke's capital expenditures by about $150 million through 1975 and result in a total capital reduction of almost $1.5 billion through 1979.

Operating cutbacks also brought some relief to the financially strapped company. Over 1,200 employees and 311 contract workers were furloughed, and the Lake Hickory Training Center was temporarily closed. The company eliminated all advertising and ended the publication of the *Duke Power Magazine*, a glossy, readable monthly designed for the general public, as well as other printed items. All company-sponsored parties and picnics were suspended, and new-car purchases deferred.

As emergency measures to raise cash, Duke sold two million pounds of "yellowcake," which is partially refined nuclear fuel in a powder form; all of its combustion turbines; and its new office building that was under construction in Charlotte. The nuclear fuel, the combustion turbines, and the building were leased back after their sale, but Horn explained that the sale reduced the amounts that the company would have to seek in the securities market and would "be of short-term help."[2]

For long-term help, Duke Power pragmatically swallowed its ideological pride and agreed to do something that it had earlier strongly frowned upon. Publicity concerning the company's financial difficulties in late 1974 inspired representatives of a group of North Carolina municipalities and REA cooperatives, both of which groups purchased bulk power from Duke Power, to inquire if the company would have any interest in selling a portion of the projected Catawba Nuclear Plant to them.

In 1967 Duke Power had successfully fought off an attempt by a group of North Carolina municipalities to acquire an ownership interest in the Oconee Nuclear Station. By the mid-1970s, however, not only had economic circumstances drastically changed, but also the entire electric industry was entering an unprecedented and strangely different era from the 1960s.

Initially, Duke took a cautious stance concerning the possible sale of a portion of the Catawba plant. True, the municipalities could finance their share of the plant by selling tax-exempt revenue bonds, a type of financing not available to Duke Power. And the REA cooperatives could get loans backed by the Federal government at an interest rate 2 to 4 percent below what Duke would have to pay for first mortgage bonds. Since the Catawba plant had to be built to avoid power shortages in the 1980s, Duke's wholesale customers were, in reality, proposing an alternate and more economical (for Duke Power) way of financing the plant.

Horn and other Duke managers involved in the negotiations candidly explained to the representatives of the municipalities and cooperatives that if the market price of Duke's common stock rose substantially above its book value—and showed promise of staying there—Duke would not be interested in selling any portion of the Catawba plant, for then Duke could raise whatever capital it needed. In August 1974, however, when the company announced the postponement of Catawba and the other generating units, the market price of Duke Power common stock was $10 to $11 and its book value around $20 per share. As of December 31, 1975, the picture had brightened somewhat, for while the common stock's book value was $19.18 per share, the market price (thanks to a rate increase Duke had won) had risen to around $19.[3] Although the market price finally did rise (briefly) above the book value in 1976, that turned out to be a temporary phenomenon, for double-digit inflation roared back with a vengeance in the late 1970s, and the price of Duke's common stock again fell below its book value.

The negotiations were protracted and complex, but by late 1978 the sale of 75 percent of the Catawba Nuclear Station's Unit 2 (which was still under construction) to the North Carolina Municipal Power Agency Number 1 was effected. Duke Power received an initial cash payment of $278.4 million, which amount was based on the costs already incurred by the company, plus a reasonable profit and fees that had been agreed upon. As Bill Grigg explained, the cash payment Duke received was applied to the company's other construction, thereby reducing outside financing requirements. Moreover, the sale of that 75 percent portion of the plant reduced Duke's construction budget, again relieving the company's external financing requirement. Under the terms of the deal, Duke Power would complete the necessary construction, operate the plant, and purchase any excess power that the municipal agency might not need.[4]

By 1981 continuing negotiations about the sale of Catawba resulted in the sale of the remaining 25 percent of Unit 2 to a group of South Carolina municipalities. And in the same year Duke sold a total of 75 percent of Catawba's Unit 1 to two electric membership cooperatives composed of Duke's North Carolina and South Carolina REA cooperative customers. Duke retained a 25% ownership of Catawba Unit 1.

The sale of most of a nuclear plant to two categories of Federally-favored customers was unorthodox, certainly, but it was one strategy for climbing out of a deep financial hole. The negotiations concerning just the initial sale of 75 percent of Unit 2 to the North Carolina municipalities spanned over three years and required forty formal meetings between the principals as well as countless informal conferences.

Douglas W. Booth, then Duke Power's senior vice president for retail operations, led Duke's negotiating team. Booth was an engineer by training, but he displayed a keen understanding of the company's financial plight. Some engineers, like some members of university faculties, are so preoccupied with their own professional specialties and interests that they remain indifferent to or uninformed about such mundane matters as the financial bottom line. Booth, however, proved able to see the broader picture, and for his patience and perseverance in the negotiations for the sale of the 75 percent of Catawba Unit 2, in 1979 he received one of the company's coveted Robinson Awards.

Begun in 1961, the W. S. O'B. Robinson Awards were named in honor of a long-time general counsel of the company in the early days who also had served as chairman of the board of directors. Each year awards went to three employees who had been nominated by their fellow employees and then selected by a committee of employees. By 1978, 51 Robinson Awards had been given: 24 for improvements in design and procedure; 16 for life-saving and human relations; and 3 for meritorious service to Duke Power in emergency situations. One of the winners in 1978, for example, received the award for a design improvement at Duke's nuclear facilities that it was estimated would eventually save the company $300 million over the lives of the nuclear units. Doug Booth, therefore, joined a select group when he won his Robinson Award in 1979.[5]

In addition to the sale of most of the Catawba plant, another important strategy for the long-range improvement of the company's financial health took the form of a radical new approach to the whole idea of selling and using electricity. With the output of each new generating unit that had to be added to the system costing more than that of its predecessor, which was the reverse of the situation in the 1950s and 1960s, a new goal for Duke Power came to be the avoidance, as much as that was possible, of having to build new generating units and plants.

After World War II, peak loads on the system rose dramatically each year as more and more businesses and homes turned to air-conditioning, electric heating, and other electric-power-consuming systems and devices. Having always to plan ahead as carefully as possible to have production margins that stayed ahead of the peak loads, in order to avoid shortages and power blackouts, Duke had nevertheless always pushed hard for the sale of more and more electricity. And building the plants to meet the ever-rising demand had been a straight-forward, manageable matter. Now, however, in the inflation-ravaged 1970s, Duke's managers began to realize that since they wanted to build as few new generating units as possible, they had to re-think the whole business of selling and using electricity.

Retired, veteran officers of Duke Power looking back from 1997 agreed that one of the major culture changes in the company's history occurred in the mid-1970s when Duke Power abandoned the "sell, sell, sell" marketing strategy of the 1950s and 1960s. "The company started spending more money to tell you how not to use electricity than it did to [tell you how to] use electricity," B. B. Parker noted, "and to me that is a whole philosophical change that I think has permeated the industry and has certainly guided Duke Power...."[6]

A key instrument in the philosophical change mentioned by Parker was a new approach to and understanding of "load management," which had as its primary objective the reduction of the rate of growth of peak demand for electricity in the Duke Power service area. The initial goal of the program, as spelled out in 1975, was to reduce the growth in peak demand by 1.3 million kilowatts by 1990.

In addition to inflation as an explanation for the rising costs of new plants, Doug Booth pointed out that large capital expenditures were now required because of environmental considerations. Since those expenditures did not produce any additional electricity, customers had to pay higher rates without receiving increased benefits, at least as far as electrical service alone was concerned. To illustrate his point about rising costs, Booth noted that Oconee Nuclear Station had been completed in 1974 at a cost of $179 per kilowatt. McGuire Nuclear Station, on the other hand, was expected to be completed in 1980 (which it would not be) at an estimated $398 per kilowatt. (The cost actually ended up being a bit over $780 per kilowatt.) While all electric utilities in growth situations were experiencing the same problem, Duke was actually better off in this regard since it designed and built its own plants at a lower cost than most other utilities.

Why did not Duke just stop altogether the building of new plants? Because, Booth explained, "a certain amount of new generating capability is absolutely essential to the well-being of the people we serve." If people were to improve their standard of living, more and more better-paying jobs had to be provided. All of those jobs would require some form of energy, and with other forms of energy (such as oil and gas) already in short supply, it was "inevitable that electricity will have to play an even bigger role in the future in providing jobs for people and in producing the goods and services that people will require." So a certain number of new plants would have to be added periodically to replace worn-out facilities, if for no other reason.

"The purpose of Load Management," Booth explained, "is not to restrict the availability of power..., but rather to help people use the electricity they need in such a way as to minimize the amount of more expensive new generating capability that will be required to provide that power."

Turning to the specific plans for accomplishing the goal of slowing the growth of the peak demand, Booth noted that Duke Power already had twenty-one separate activities underway with that purpose in mind, and there would be other approaches later. One category included programs that encouraged customers to shift certain uses of electricity from on-peak to off-peak times. Another was to help customers improve the efficiency of their on-peak uses of electricity that could not be shifted. And a third category involved the possible voluntary control of certain electric devices used by customers during peak demand periods.

Washing machines, clothes dryers, and dishwashing machines were examples of devices, which, in most cases, could be used at virtually any time of day. In theory, shifting non-time-sensitive uses of electricity from peak to off-peak was an ideal marketing concept. It not only reduced the amount of generating capability required, but it also increased the output of generating plants at times when otherwise much of the capacity would be idle. Those economies of operation not only helped hold down the cost of electricity, but they also enhanced the company's prospects of actually earning the return on its shareholders' investments that regulatory agencies had determined to be fair and reasonable. Booth believed that some, certainly not all, customers would voluntarily change certain living habits, but the company was going to have to do a lot of educational work and offer pricing incentives.

According to Booth, Duke was much interested also in the concept of time-of-day rates, also known as peak-load pricing. The company had asked the North Carolina Utilities Commission for permission to put time-of-day rates into effect on a voluntary, experimental basis involving a limited number of customers. Since there were also some significant costs associated with the time-of-day rates — they would require special meters, for example — the benefits would have to outweigh the costs for the plan to be viable.

As for improving the efficiency of on-peak usage, more needed to be done, even though in the wake of the Arab oil embargo people were already trying to conserve energy. Instead of expecting people to turn off their air conditioners on hot days, Duke was pushing a program that would allow them to run air conditioners but use less energy in the process. Inadequate levels of insulation in homes caused equipment for heating and cooling to work much harder and use more electricity. Duke, therefore, had developed a set of standards for what it called an Energy Efficient Structure. The standards required far greater levels of insulation than did the states' building codes, and homes qualifying for the rate assigned for an Energy Efficient Structure had to have double-paned glass or storm windows, insulated or storm doors, and, in the case of new homes, a limit on glass of 12 percent of the outside wall area.

Duke's goal was to have about 15,000 homes in its service area meeting those standards by 1980 and about 220,000 by 1990. It was promoting the program through an extensive consumer-education program aimed not only at homeowners but also at builders, architects, and others in the home-building and home-improvement fields.

Another energy-conservation program pushed by Duke aimed at improving the efficiency of major appliances such as water heaters, refrigerators, and air conditioners. The company was encouraging manufacturers to build more energy-efficient models, dealers to stock them, and consumers to buy them.

As for using certain devices to control customer demand during peak periods, Duke was studying their feasibility. In addition to having sufficient generating capability to meet peak demand, a power company had to have a certain amount of generating reserve to insure the reliability of the system. The essential margin or reserve could be provided in either one of two ways. The conventional way was to have actual generating capability ranging from 15 percent to 25 percent above expected peak demand. The other way was to be able to automatically reduce the demand, if necessary, during the peak-demand period. Obviously, the advantage of the second way was that the company could provide the required reserve margin with less actual generating capability and at less cost.

One way to accomplish that would be to have an interruptible rate agreement with some of the company's large industrial customers. Duke would be allowed to reduce rather sizable loads but of the type that would not materially affect the customer's operation during the period of interruption. Another example of how demand could be reduced would be the interruption of service to certain home appliances such as water heaters. In most cases, the water in the tank would remain heated for the length of time that service would be interrupted.

Booth conceded that with the various programs that he described the company expected some reduction in the growth of kilowatt hour sales. But there would be a proportionately greater reduction in the growth of the peak and the amount of generating capability required to serve that peak. That would result in a gradually improving load factor, and the company also believed that shareholders would benefit by the fact that it would reduce the erosion of earnings that normally occurred between the time plant construction costs were incurred and the time they were recovered through rate increases.

In view of what was known about the costs of plants scheduled for service through the 1980s, Booth concluded, Duke Power believed that Load Management would be a viable and important marketing concept at least that long and probably even longer.[7]

Booth—and Duke Power—proved to be absolutely correct. By as early as 1977, that the Load Management program was well on target became clear. The program had aimed for a reduction in peak demand in the summer of 1977 of 206,000 kilowatts; the documented reduction through that period was 208,000 kilowatts.[8] Subsequent years showed progressively larger reductions, and in 1977, in light of the company's success with the various aspects of its load management program, the targeted goal was significantly raised: instead of aiming at a reduction of 1,300,000 kilowatts in peak demand by 1990, the revised goal was 4,508,000 kilowatts by 1994. That would be the equivalent of the output of nearly four nuclear-fired generating units.[9]

Duke Power's load management program became a model for other electric utility companies across the nation, and by 1980 a prominent Wall Street electric utility analyst hailed Duke's program as "the most aggressive we have seen in both breadth and depth of scope."[10]

Another strategy for climbing out of the financial hole consisted of Duke Power's ten-year financial plan. As proposed by Horn, Grigg, and other senior managers and approved by the company's board of directors in late 1975, the plan's purpose was to improve the company's financial strength and its flexibility in financing future plant construction. The desired flexibility could be attained, according to the plan, primarily through higher credit ratings on Duke's securities and improved liquidity. As target goals, the plan called for reducing the debt component of the company's capital structure to a maximum of 52 percent, increasing the earnings coverage of fixed charges (i.e., bond interest) to a minimum of three times, and generating at least 40 percent of total capital requirements from internal sources.[11] For the first time, the plan put an absolute limit on what the company could spend for construction.

Fortunately for Duke Power, it promptly made progress toward meeting the goals of the financial plan. In 1976 the debt component of the capital structure was reduced from 55.4 percent to 53.9 percent; earnings coverage of fixed charges rose from 2.19 times to 2.8 times; and approximately 62 percent of capital requirements during the year came from internal sources. Consequently, Standard and Poor's upgraded the rating on Duke's commercial paper from A-2 to A-1, the highest rating, and the rating on Duke's preferred stock from BBB to A.[12]

Although Duke Power still had some distance to go before regaining truly robust financial health, the company had made sufficient progress in 1976 and 1977 to be chosen by *Electric Light and Power* magazine as the "Outstanding Electric Utility for 1977." According to the trade journal, Duke won the honor "on the strength of the exceptional manner in which it has adjusted to today's more stringent operating requirements while still maintaining high-quality service throughout its area." In making the award at a luncheon in Charlotte, the editor of *Electric Light and Power* particularly cited Duke Power's strong financial recovery from the recession of 1974–75, its efficiency record in plant operation, and its leadership in the development of new technologies in both electrical transmission and distribution. The selection of Duke had been made, he noted, by a panel of experts representing the electric utility industry, the manufacturers of electrical equipment, and industry analysts.[13]

Another honor came to Duke Power earlier in 1977 when *Financial World* named Carl Horn, Jr., as the electric utility industry's "outstanding chief executive officer." The company's board of directors indicated its own appreciation of Horn's leadership through the economic storms of the early 1970s by naming him in 1976 as the company's first chief executive officer and chairman of the board of directors. At the same time, B. B. Parker was promoted from his post as executive vice president and general manager to that of president and chief operating officer. When Parker retired two years later in 1978, Bill Lee was named to succeed him as president and chief operating officer.

B. B. Parker, who went to work for Duke Power in 1936, was a forty-two-year veteran. Such extended service was by no means uncommon at Duke, and the *Duke Power News*, a tabloid-sized, monthly publication for employees which began appearing in 1975, devoted much of its space to pictures and short biographical sketches of all retirees as well as of those who had been with Duke Power for twenty-five, thirty, or thirty-five years. Parker represented top management, of course, but one blue-collar Duke employee who retired in 1975 may be used to at least suggest certain attributes of many Duke employees as well as, one suspects, of many other industrial workers in the Piedmont Carolinas, especially in the first five or so decades of the century.

Roy Edwards had gone to work for Duke Power in 1944, so his service with the company was a decade shorter than Parker's. Edwards worked on the transmission lines, his first job involving the painting of towers at 40 cents per hour. He formed the habit of getting up at 2:30 or 3:30 a.m. because he normally left home at 5:00 a.m. to be on the job at 6:00 a.m., an hour before the regular work day began. "I'd rather be an hour early than 5 minutes late," Edwards explained.

He had been a farmer prior to getting the job with Duke, and the last year he farmed, he noted, "I couldn't even pay my fertilizer bill. Those were hard times." He had, at one period in his life, walked 14 miles a day to cut wood for 75 cents a cord.

When he retired as a Duke lineman, he was overseeing the work of contract crews who cleared the right-of-way for transmission lines. "They [Duke Power] did me a favor when I first went to work for them," Edwards declared. "At first I was afraid because I didn't have an education, but I can say one thing—Duke Power sure has been good to me." A member of Union Grove Baptist Church, he had been married for 46 years but, he added diplomatically, "It don't seem like it." He and his wife had 10 children, 24 grandchildren, and 4 great-grandchildren.[14]

Not all Duke Power employees shared Roy Edward's work ethic, of course. Many of them probably did, however, for they too came from impoverished rural backgrounds, especially in the first fifty or so years of the company's history. Southern agriculture slumped badly in the 1920s only to be knocked down even further during the Great Depression of the 1930s. A regular-paying job, such as Edwards landed in 1944, was not just a boon but a lifesaver. He regarded it accordingly and probably raised his children similarly to appreciate steady employment.

Bill McGuire, Carl Horn, Bill Lee, Bill Grigg—all insisted that Duke Power's greatest strength lay in its employees. For example, in his speech to the annual stockholders' meeting in 1985, Bill Lee finally had a lot of good financial news. He insisted, however, that it was the Duke employees, "21,000 dedicated, talented men and women—who make Duke Power what it is."[15]

An additional point along this line might be hazarded. While Duke Power itself played a significant role in the economic transformation of the Piedmont Carolinas in the twentieth century, one has to suspect that it was the region's people themselves who played an even larger and more fundamental role in that transformation—the thousands and thousands of Roy Edwardses who wanted to work and did so, at least most of the time, willingly and conscientiously.

Confirmation that Duke Power was on track as it strove to regain its financial strength came from an impartial, outside source in 1976. The North Carolina Utilities Commission ordered that all electric utilities operating in the state undergo a management performance audit to provide an impartial, professional assessment. Accordingly, after an eight-month study of Duke Power, Booz, Allen and Hamilton, Inc., a nationally recognized management consulting firm, reported that it had found Duke to be "a well-managed company operating generally in a cost-effective manner in all major functional areas."

In the consulting firm's summary comments, it reported that, "Duke has clearly demonstrated during the course of this audit why it is considered a leader in the electric utility industry: the company conducts its affairs in a generally outstanding manner." The audit noted "particular strengths" in the following areas: Duke had developed "a strong, experienced management team"; managers at all levels were "both technically competent and cost conscious"; management exhibited "pride in past accomplishments but also a positive attitude toward constructive changes"; a technological leader in the electric utility industry, Duke "consistently ranks at or near the top of comparable utilities in the relative efficiency of its generating plant design, construction, and operating practices"; the company made "extensive use of state-of-the-art applications of data processing and telecommunications technology..."; and major expense items—such as fuel, construction costs, and interests expense—were "tightly controlled." The audit report concluded by noting that opportunities for improvements in managerial and operating efficiency, in most cases, represented refinements of existing practices.[16]

No doubt enheartened by such a positive assessment from a prominent consulting firm, Duke Power in the late 1970s took additional steps to change and strengthen itself. The company's board of directors had traditionally consisted of its senior officers and certain trustees of the Duke Endowment; that was precisely the close, interlocking relationship between the power company and the Endowment that J. B. Duke had wanted.

With that type of close tie between a profit-seeking business and a tax-free philanthropic foundation under attack in the 1960s, however, Bill McGuire persuaded the board of directors that the time was right for the election of outside directors to the board. Accordingly, in 1966 McGuire, seeking one new director from North Carolina and one from South Carolina, persuaded Howard Holderness, then chairman of the board of Jefferson Standard Life Insurance Company and Jefferson-Pilot Corporation of Greensboro, North Carolina, and Dr. Robert C. Edwards, then president of Clemson University in South Carolina, to join the board.

In the mid-1970s Carl Horn and the board gradually moved to have the majority of the directors come from outside the company. In 1975, the board elected the first woman and the first African American as members—Dr. Naomi G. Albanese, dean of the School of Home Economics, University of North Carolina at Greensboro, and John S. Stewart, president of the Mutual Savings and Loan Association of Durham, North Carolina.

Still seeking the expanded expertise and experience that outside directors could bring to the board and also following a policy favored by the influential Wall Street analysts, Duke's board brought in additional new members from outside in 1976. By the middle of that year, therefore, there were nine outside directors and only seven company officers on the board.

Along with the addition of the outside directors, Duke moved aggressively in the 1970s to increase the number of women and African Americans holding non-traditional positions as well as doing business with the company. In the latter category, for example, Duke moved $26 million of its $260 million in employee life insurance coverage from the company with which Duke had long dealt, the Pilot Life Insurance Company of Greensboro, North Carolina, to the North Carolina Mutual Company of Durham, North Carolina, the nation's largest insurance company run by African Americans.[17]

The first woman engineer in construction, Heidi Valenta, began working at the construction site of the Catawba nuclear plant in 1976. Asked about her reception from her male co-workers, she declared, "They're nice to me like they would be to any girl who paints her fingernails."[18]At the construction site of the Cherokee nuclear plant, three women (not engineers) were driving utility trucks and a 50-ton roller. All three reported positively on their experiences, with one of them adding this comment: "I like to be treated like a woman. I like the same pay—but not necessarily to be treated like a man because of it."[19] Women were taking on all sorts of non-traditional jobs with Duke. The first woman welder to be trained and certified by the company, Willie Mae Eades, had spent sixteen years in a hosiery mill before joining Duke Power, where she began as a construction worker. "I decided that if I was going to stay with Duke," she recalled, "I wanted to get into something where I could make some money."[20]

The first woman to work on a Duke line crew, Pat Wagner, came to the job fresh out of the United States Air Force. She had been urged to apply for the job by a male Duke employee who knew that Duke was trying to recruit women for non-traditional positions. Asked if the guys on the crew hassled her, she said, "No, not really. They pick on me, but it's friendly." The only reservation one male crew member had about her was that she did not play golf, fish, or chew tobacco.[21]

By 1982 Duke had its first female nuclear-plant control room operator, one of only about a dozen female control room operators in the country.

There were also reported to be several more women technicians who were going through Duke's elaborate and extensive training program in preparation for taking the stiff examinations given by the Nuclear Regulatory Commission to those who wished to be certified as control room operators.

That women should move into all types of work connected with Duke's nuclear plants was only natural, for few electric utilities in the country had made a larger or more enthusiastic commitment to nuclear-fired generation than had Duke Power. With Oconee Nuclear Station's three units in operation since 1973–1974 (unit 1 in 1973, units 2 and 3 in 1974), Duke in the late 1970s had four additional nuclear plants in the works: McGuire would begin commercial operation of Unit 1 in 1981 and of Unit 2 in 1984; Catawba Unit 1 began commercial operation in 1985 and Catawba Unit 2 in 1986; and as for the famed "Six-Pack" so boldly announced in 1974, site construction had begun at Cherokee but not at Perkins.

The actual performance of Oconee Nuclear Station gave Duke Power every reason to be enthusiastic about nuclear-fired generation. By 1975 Oconee, then the world's largest operating nuclear station, generated 15.3 billion kilowatt hours of electricity, which represented about one-third of Duke's total generation for the year.[22] By 1979, Duke estimated that Oconee's three units had saved the company's customers over $590 million in fuel costs. Despite the higher capital costs associated with nuclear generation (because of construction costs), the station's cumulative savings to customers had nearly equaled Oconee's original construction cost.[23]
W. O. Parker, Jr., Duke's vice president for steam production, explained about savings made possible by Oconee. "Every day we operate the plant at full capacity," Parker stated, "we save the equivalent of 27,000 tons of coal." To put the matter another way, Parker noted that it cost about 11 mills (1.1 cents) to generate one kilowatt hour of electricity in coal-fired plants, while the cost per kilowatt hour at Oconee, excluding capital costs, was 3 mills (0.3 cent).[24]

By 1979 Oconee had achieved over 3 million manhours without a disabling injury. Duke understandably took great pride in that, for no other nuclear plant in the country and no other Duke Power installation had ever achieved such a long safety record.[25]

Despite such a record, there were, of course, highly vocal anti-nuclear groups all over the nation, including the Carolinas. In 1977 one of those anti-nuclear groups opposing Duke's construction of the McGuire and Catawba plants, won the first round of a legal battle that was crucial for the nuclear power industry. In 1957 Congress had passed and President Eisenhower had signed the Price-Anderson Act to provide supplemental nuclear liability insurance for the electric utilities. Price-Anderson limited a utility's liability for damage resulting from a single nuclear accident to $560 million, the total amount of coverage then available from both private and government sources.

The members of the anti-nuclear group, claiming that Price-Anderson was unconstitutional because it allegedly violated their rights of due process and equal protection under the law, sued in the Federal district court in Charlotte, and the judge there ruled in favor of the plaintiffs, the anti-nuclear group.[26]

Duke Power promptly appealed that ruling, and in 1978, in an unanimous decision, the United States Supreme Court overturned the ruling of the district court judge and declared that Price-Anderson was "an acceptable method for Congress to utilize in encouraging the private development of electric energy by atomic power."[27] (In 1988 President Reagan signed into law a fifteen-year extension of Price-Anderson, which legislation also raised the liability limit for damages from a single nuclear accident to about $7 billion. Although the new act substantially increased an individual company's exposure to liability in a worst-case accident, it also assured the electric utilities of a stable nuclear-insurance environment into the next century.[28]

No one in Duke Power was higher on nuclear-fired generation than Bill Lee, who became the company's president and chief operating officer in 1978 (with Carl Horn still serving as chief executive officer and chairman of the board of directors). Lee had acquired, in fact, an industry-wide reputation as both brilliantly knowledgeable and highly articulate about the advantages of nuclear plants. Even when the fate of the Price-Anderson Act was still undetermined, Lee declared: "Nuclear's track record has clearly demonstrated two things. First, that it is the lowest cost type of generation available...; secondly, that it has an outstanding safety record. Even without a liability limitation, we expect to continue with our nuclear program because of those two factors."[29]

The American public at large probably did not concern itself so much about nuclear's cost advantage over coal. And while some environmentalists conceded nuclear's relative environmental attractiveness as far as the air was concerned, other strongly objected to nuclear plants. For the general public, the most convincing and reassuring claim concerning nuclear-fired generation was that, as the *Duke Power News* declared, nuclear power had "a demonstrated safety record second to no other industry....."[30] That was the proud boast that came crashing down in March 1979.

A cooling system malfunction that occurred on March 28, 1979, at a nuclear plant in Pennsylvania known as Three Mile Island was not, in and of itself, all that unusual. In responding to the problem, however, the operators blundered—and turned a problem into an extremely frightening and potentially dangerous crisis. One of the more rational objectives to nuclear power had long been that if the cooling system malfunctioned, that could precipitate a meltdown of the nuclear fuel, with a possible explosion that might shatter the containment structure and send radioactivity into the surrounding environment for many miles.[31]

The containment structure of the Three Mile Island nuclear unit remained intact during and after the accident, and no one was killed. Avoiding the dreaded meltdown, however, was a touch-and-go affair for many hours, and for days erroneous and exaggerated reports of core meltdown, escape of radiation, and potential explosion frightened the American people, especially in Pennsylvania and all neighboring states. The bottom line was that a serious nuclear-plant accident had occurred, and there could be no more proud assertions about nuclear plants' "unblemished safety record."

Immediately upon learning of the accident at Three Mile Island, Bill Lee and a carefully hand-picked group of Duke employees rushed there. They joined a team of over 2,000 scientists, engineers, and others who worked successfully to bring the overheated unit under control. As already stated, there were no deaths, no meltdown, no escaping radiation—but it was a hair-raising close call.

Soon after the incident, Bill Lee and other Duke Power experts in nuclear power met in New York with a group of utility analysts. The purpose of the meeting was for Lee and his colleagues to report on the accident at Three Mile Island and to explain and answer questions about Duke's involvement with nuclear-powered facilities. One of Duke's experts stated that the problem at Three Mile Island stemmed basically from "equipment malfunctions" compounded by "human error." Duke Power, it turned out, had units similar to those at the Pennsylvania facility and had encountered most of the operational problems that had occurred at Three Mile Island, but never in sequence.

In light of the incident, Duke Power had modified its operational procedures and equipment. Although Duke's basic operational methods were different from those at the Pennsylvania facility, Duke was reassessing all of its reactors and methods, and all operating personnel were undergoing resimulation training. National authorities had reviewed Duke's plants and operations but had found no problems. The Federal licensing process for nuclear facilities, nevertheless, would be slowed down, and two of Duke Power's projected nuclear plants (Cherokee and Perkins) would be further delayed. "Generally the analysts . . . seemed favorably impressed by Duke's presentation," the report concluded.[32]

At Duke Power, Lee spoke for the company when he insisted that, despite Three Mile Island, "nuclear power was still the best energy option" and that, despite "controversy and political uncertainties," Duke remained committed to it as the "safest, cleanest and most economical source of electric energy." For the electric utilities, generation by nuclear and coal was the "only viable alternative to increased dependence on foreign oil."

Lee admitted that there were many obstacles to be overcome. The biggest, he argued, was the continued absence of a strong national energy policy which recognized that the future of the country rested to a certain extent with the nuclear option. Without that recognition, there was inadequate incentive within the Federal bureaucracy to make timely decisions on the licensing of nuclear plants, the reprocessing of spent nuclear fuels, and the ultimate disposal of nuclear waste. All those issues were critical to increased utilization of nuclear power. While the technology existed to resolve those issues, the incentive to do so was dampened by the Federal regulatory process. Lee asserted that until the Federal administration and Congress formally recognized that the United States could not have an adequate energy supply without nuclear power, there would continue to be regulatory impediments.

Lee admitted that many of the wounds suffered by the nuclear power industry had been self-inflicted. The Three Mile Island accident, along with a number of less serious problems that had been publicized, had eroded public confidence. To gain the degree of public acceptance necessary to remove the political obstacles, the nuclear power industry had to convince the American people that nuclear plants could be operated with lower risk than any alternative energy source currently available.

Among the many lessons learned from Three Mile Island, Lee mentioned first the fact that meeting the minimum standards of the Nuclear Regulatory Commission was "not synonymous with safe operations." The Pennsylvania accident had resulted largely from human errors that could have been avoided through more thorough analysis of preceding incidents and more effective management and training systems.

Another lesson was that public acceptance of nuclear power did not rest with how well one company might perform with regard to safety, but how well the industry performed as a whole. "In regard to the public's perception of nuclear safety," Lee asserted, "this industry is no stronger than its weakest member."

With those lessons learned, the nuclear power industry had taken steps to build public confidence in nuclear power and to give assurance that every effort was being made to avoid serious accidents. The steps were being taken with the recognition that "regulation provides only a starting point for safety." The ultimate responsibility lay with the various companies that operated nuclear plants.

The nuclear power industry had quickly created two new safety-oriented organizations. One was the Nuclear Safety Analysis Center established in May 1979, at Palo Alto, California. It initially developed the authentic, detailed analysis of what actually happened at Three Mile Island and what the accident implied in terms of generic safety issues and remedies. Its future task would be to identify and evaluate all generic issues concerning nuclear safety, to recommend remedies, and to assure a direct line of technical communication within the industry. Thus, each utility would be able to take advantage of the total industry's expertise.

The other new organization inspired by the Three Mile Island accident was the Institute of Nuclear Power Operations (INPO). It was the brainchild of Bill Lee, who also became the first chairman of its board of directors, and while it was to work closely with the Nuclear Safety Analysis Center, INPO was focused primarily on the human factor involved in plant operations and developed elaborate educational and training programs for the operating and management personnel of nuclear plants. The organization promptly set about accrediting existing programs that met its criteria and certifying instructors. With Duke Power so prominently committed to "providing leadership within the nuclear industry," the goals were to help achieve energy independence for the United States, electricity at reasonable costs, and public safety [33]

Bill Lee took a leadership role not only within the nuclear power industry but also before the general public. The day after Jane Fonda and Tom Hayden, two prominent anti-nuclear activists appeared on "Good Morning America," a popular news and talk show on television, Bill Lee was interviewed on the show to present the pro-nuclear case.[34]

To better equip plant operators, chemists, health physicists, and maintenance personnel with technical expertise, Duke Power completed building in 1980 one of the most modern and well-equipped training centers in the nation. McGuire Technical Training Center on the shores of Lake Norman featured an array of sophisticated equipment and laboratories in addition to traditional classrooms. In a very real sense, the McGuire program was comparable to a technical college, both in its campus setting and highly structured curriculum.

The classroom instruction was reinforced with practical hands-on experience in the center's chemistry and health physics laboratories as well as mechanical and electro-pneumatic workshops. Future nuclear operators were exposed to actual control room conditions in a $3 million simulator programmed to test the trainees' skills in handling different malfunctions. A staff of 46 professional instructors gave the 150 trainees who went through the program each year an education in power plant operations that was the equivalent of a college major in certain areas; and while quality was carefully maintained, the cost per trainee was significantly less than that which would have been incurred through the use of outside training.

The McGuire facility quickly garnered widespread acclaim as a model for the electric utility and nuclear power industries. It also hosted numerous touring groups from other electric utilities, governmental agencies, and foreign countries. Further concentration of training and scientific facilities on the Lake Norman campus was continuing with the construction of a physical sciences building that was designed to hold all of Duke Power's environmental laboratories. [35]

Insisting on a positive, even up-beat response to Three Mile Island, Duke Power also celebrated its 75th anniversary in 1979. Because of the colorful and highly distinctive manner in which the company had been launched in the early Twentieth century, its leaders had always emphasized and hailed the company's history. Both in speeches before various types of audiences ranging from the Newcomen Society to professional organizations, as well as in printed form, Duke Power's successive presidents and others had recounted the story of how Dr. Gill Wylie and William S. Lee I had recruited James B. Duke and his millions, first for the development of the Catawba River and then for the gradual shift from hydro to steam power.

For the 75th anniversary Duke Power employed a free-lance writer with extensive experience in journalism and public relations, Joe Maynor, to write a series of lively articles, richly illustrated with vintage photographs covering the history of the company from 1904 down through Three Mile Island in early 1979. After the articles appeared monthly in the *Duke Power News*, they were assembled into a handsomely illustrated volume, *Duke Power: The First 75 Years.*[36]

From its earliest days, the company's leaders had kept a rich variety of documents in a carefully maintained "Central File." In 1986, however, Duke Power hired an able young professionally-trained archivist, Dennis Lawson, and he promptly set about establishing a first-rate archival and records-maintenance program.

By the time of Duke's 75th anniversary, the company had made considerable progress toward achieving the financial goals that had been set in 1975. In other words, Duke Power had climbed steadily out of the deep and scary financial hole in which it had found itself in 1974–75. The capital structure then was comprised of 30 percent common stock, 14 percent preferred, and 56 percent debt. By 1979 it had been strengthened to the point where debt had fallen below 50 percent of the capitalization and common equity (stock) had risen above 36 percent. Earnings coverage of fixed charges had been increased from slightly over 2 times in 1974 to about 3 times in 1979.

There remained problems: internal cash generation remained at an inadequate level of about 25 to 30 percent of total requirements. That was expected to improve as construction work in progress got factored into the rate base, pursuant to new legislation in North Carolina. Both earnings per share and dividends had increased, but the market price per share of common stock remained below book value. For that reason, therefore, Duke felt forced to delay further the completion of Cherokee Nuclear Station rather than accept the further dilution of book value per share which would have resulted from the sale of common stock that would have been required by the old schedule for the construction of Cherokee.[37]

Despite the company's multifaceted campaign to strengthen itself financially—the load management program, the sale of most of the Catawba nuclear plant, plant deferrals, and cost-cutting measures—fundamental weaknesses in the economy at large handicapped Duke Power as well as other electric utilities and business in general. The most basic problem was that double-digit inflation came rampaging back in the late 1970s only to be followed by a sharp recession in the early 1980s. Carl Horn insisted that Duke Power had worked diligently to soften the "impacts of inflation and excessive regulatory requirements." The company had guarded against unnecessary expenses, vigorously sought rate relief as necessary, and sold assets to relieve financing pressure. In dealing with the symptoms of inflation, Horn maintained that Duke had done its job well. Then, speaking candidly about political matters, he added: "But the principal causes of inflation—deficit spending by the federal government and excessive regulations which add billions of dollars to the costs of providing goods and services—remain unchecked."

President Jimmy Carter had appealed for wage and price restraints in what was sometimes termed a "jawboning" approach to the problem of inflation. Horn stated that Duke would comply with the president's request but felt disappointed with that approach to the problem for two reasons. "First, it places the burden of fighting inflation on those sectors of our society which have been least responsible for creating inflation, but [which] have been inflation's victims." Secondly, wage and price controls were simply another form of treating the symptoms of inflation while doing nothing to overcome its underlying causes. "The President and Congress could best serve the people," Horn argued, "by imposing restraints on federal spending, removing unnecessary regulatory impediments to higher productivity, and encouraging investment in industry."

Horn thought he discerned signs on the economic horizon of what appeared to be "a growing public demand for fiscal responsibility by government at all levels.... He could only hope that such thinking would continue to flourish and result eventually in a balanced budget and "a relaxation of federal reins on the American free market system."[38] The balanced Federal budget that Horn so coveted would not be realized until the 1990s, of course, but long before that happened, Duke Power set about trying to broaden its financial base and add to the bottom line.

As a pilot project in diversification, Duke undertook to design a waste facility for another utility. Duke Power's resources for diversification included expertise in engineering and construction, as well as two wholly-owned subsidiaries, Crescent Land and Timber Corporation and Mill-Power Supply Company. Approaching diversification with caution, Duke at that point did not expect its unregulated activities to constitute a significant portion of operations in the foreseeable future. "As opportunities arise," Carl Horn and Bill Lee advised the shareholders in 1981, "we will pursue them prudently while working to maintain our high standards of electric service." [39]

Subsequent years would bring more diversification than Horn and Lee envisioned in 1981, but Carl Horn would not be involved. *Financial World* magazine again named Horn as the electric utility industry's outstanding chief executive officer in 1980, an honor that Horn had won earlier in 1977. In 1981 the *Wall Street Transcript* chimed in with a similar accolade for Horn, and perhaps those plaudits helped erase some painful memories of 1974–75.

At any rate, early in 1982 Horn announced that he was voluntarily retiring at age sixty. He commented that, while the company still faced problems, he frequently admonished his colleagues on the executive committee, "Cheer up men, we survived in 1974, so we can survive anything."[40]

To succeed Horn, the board of directors named Bill Lee as chief executive officer and chairman of the board. Aside from his nuclear expertise, Lee was perceived by friends and colleagues as having boundless energy and enthusiasm. "I do have a tendency to go throttle-wide-open," Lee confessed. "I have lots of energy. But sometimes my clock runs down like anyone else's." He admitted also that his impatience at times was a shortcoming and added: "In my haste, I sometimes hurt people unwittingly. And I feel ashamed when I do, because I'm inclined to be kind." [41]

At the same time that Lee moved up to become chief executive officer, the board named Douglas W. Booth as Duke's new president and chief operating officer.

The election of Lee and Booth to the company's top positions coincided with some unusually tough decisions that had to be made. Since 1974, Duke had remained committed to the building of the "Six-Pack," two nuclear-fired generating stations, Cherokee and Perkins, with three units at each. The construction schedule had been stretched out in the case of Cherokee and deferred in the case of Perkins, but the presumption all along had been that the plants would be built.

In 1982, however, Duke Power finally faced up to several unpleasant but inescapable facts: the forecasts for economic growth in the service area had been reduced, as had those for growth in electricity usage; there were regulatory and economic uncertainties; and, perhaps most important, there were serious difficulties about attracting the necessary capital. Consequently, on management's recommendation the board of directors voted to abandon plans for all three units at Perkins and for units 2 and 3 at Cherokee. The status of Cherokee Unit 1, which was much farther along toward completion than the other two units, remained unchanged for the time being.

"Cancellation of these units," the company explained, "will minimize the need for additional stock offerings below book value and lessen exposure to volatile capital markets." Duke planned to seek to recover through rates the costs associated with the cancelled units.[42]

Since Duke had never begun construction on Perkins Nuclear Station nor even received the necessary federal construction permits for it, the cancellation of Perkins was a relatively straight-forward and painless affair. The company expected to recover through rates the approximate $8.9 million that had been spent on preliminary engineering and licensing.

The Cherokee plant, however, was another matter altogether. While construction of its Unit 3 had not begun, that was not true of Unit 2 and even less true of Unit 1. In light of the rapidly escalating costs of nuclear plants, due partly to inflation and partly to ever-increasing and changing regulations, especially after Three Mile Island, cancellation of plans to start or complete construction of nuclear plants became commonplace across the nation. Some utilities, however, soldiered on with the building of nuclear plants that ended up being ruinously expensive.

In Duke Power's case, management and directors made the particularly painful decision in 1983 to cancel even Cherokee's Unit 1. Looking back from a later time, Steve Griffith, senior vice president and general counsel in 1983, declared, "...I can't believe that we had a bigger decision" during his more than thirty years with Duke Power than the cancellation of all units at Cherokee. Cancellation of Perkins was easy, but "we thought we needed Cherokee, and we had spent [about] $600 million on it, and it was a difficult, difficult decision...."[43]

Because the North Carolina Utilities Commission had originally agreed about the need for Cherokee and approved the plans for its construction, the Commission initially allowed the recovery of Duke's investment in the project over a ten-year period. The attorney general of North Carolina, however, strongly attacked Duke Power for its costly miscalculation, and the North Carolina Supreme Court ended up allowing Duke to recover only about a third of its costs at Cherokee in rates; Duke got a write-off on its taxes for about a third, and the shareholders paid for about a third.[44]

By obeying the First Law of Holes and ceasing to dig deeper—even though the strategy involved unorthodox steps, such as the sale of Catawba, and painful ones, such as the cancellation of Cherokee—Duke Power broke through to much brighter financial sunlight around 1983. The economy's recovery in the second half of that year helped with sales, and the dividend on common stock was increased for the eighth consecutive year. The most far-reaching development, however, was the improvement in the company's long-term financial strength. It was affected dramatically by a major reduction in the construction program and continued improvement in the capital structure. As a consequence of that, three rating agencies upgraded their credit ratings on Duke's fixed-income securities.

To avoid additional construction as long as possible, Duke was continuing to push load management and beginning a new program to increase sales during off-peak hours. "The reduced need for new construction marks a major change for Duke Power," Lee and Booth reported, "allowing us to further refine our operations and focus on new opportunities."[45]

The company had no public sale of common stock in 1983 and anticipated none in the then foreseeable future. It did, however, raise a total of $84.3 million by issuing more than 3.6 million shares through its stock purchase and dividend reinvestment plans. The company introduced its customer stock purchase plan in 1983 in an effort to broaden its investment base and increase customer understanding of issues affecting Duke Power. Under the plan, customers were eligible to purchase Duke common stock in amounts as small as $25 or as large as $3,000 per quarter without paying brokerage fees. At year-end, more than 8,000 customers were enrolled in the plan, having invested about $9.4 million in the company.

At the same time, participation in Duke's dividend reinvestment and stock purchase plan continued to grow. At the end of 1983, 33 percent of the company's common stock shareholders and 14 percent of its preferred shareholders were participating, investing an additional $28.9 million during the year.

Duke began purchasing shares of its common stock in the open market to satisfy the requirements of the employees' stock ownership plan. Also, open-market purchases for the stock-purchase savings program for employees began in early 1984. Continuing that practice meant that Duke could avoid issuing about 10 million shares of new common stock through 1987.

Along with the relatively novel encouraging financial news, Duke Power continued in the 1980s to lead in the nation for efficiency in plant operations, which was not at all novel news. By 1983 the company's nuclear units generated 42 percent of the system's total output; coal-fired plants accounted for 54 percent and hydroelectric plants for 4 percent.

Oconee Nuclear Station, ten years old in 1983, had its best operating year ever, recording a 79.2 percent capacity factor. (Capacity factor represents the portion of potential generation that is actually achieved by a facility.) Moreover, Oconee was the most efficient nuclear plant in the nation for the third consecutive year.

Duke's coal-fired plants, especially Marshall and Belews Creek, topped the national list for efficiency so regularly that the news became routine. And the company always liked to make this point in connection with the high ratings for efficiency: "Duke customers would have faced more than $90 million in additional fuel costs in 1982 had the company's generating system performed at the median level of the utilities surveyed."[46]

Another type of efficiency resulted from a Corporate Goals Program that Duke Power initiated in 1981, with the purpose of improving productivity and profitability. Employees responded to the challenge, and the opportunity to receive bonuses in Duke stock, by achieving 27 of the 31 targeted goals, and that saved Duke and its customers millions of dollars in labor, material and administrative expenses. Some examples of the results of this program were that Duke succeeded in increasing less-expensive nuclear power production by 50 percent; and the company continued to surpass its fuel efficiency goals for coal-fired plants. Other performance goals achieved during the previous three years were a 44 percent improvement in service reliability to customers; a 29 percent drop in the number of accidents involving company vehicles; and a 26 percent reduction in the number of disabling injuries on the job.

The success of the Corporate Goals Program inspired the introduction of other programs designed to reduce expense and improve employee performance. Through a cost reduction program begun in 1982, for example, employees identified ways to cut operating and capital costs by more than $10 million. Suggestions adopted ranged from switching industrial cleaning agents, for an approximate savings of $12,000 a year, to reducing the volume and cost of processing low-level radioactive waste at Oconee Nuclear Station for an estimated annual savings of $1.2 million.

Many departments at Duke were using quality circles to involve workers directly in solving day-to-day problems. Salaried employees worked under a performance management program that clearly defined performance objectives and tied raises and promotion opportunities to meeting them.

In addition to stressing day-to-day cost efficiency, Duke worked on several fronts to reduce the need for new construction and encouraged both commercial and industrial customers to install sophisticated energy management programs. At the then-new Charlotte/Douglas International Airport, for example, three 125-ton water chillers reduced the air-conditioning load during periods of peak demand. The system chilled and stored water during off-peak periods and recirculated it for cooling during on-peak hours.

Duke Power also experimented with advanced energy management technology in its own buildings and laboratories. A 40-module, zinc-bromide battery was installed in the Charlotte headquarters to store power produced off-peak for use on-peak. Such a battery might ultimately have a wide range of utility and industrial application.

As another way of minimizing the need for new construction, Duke was modernizing its older power plants in order to extend their lives. As a part of that effort, microprocessors were installed so that plant systems could be monitored continuously and more precisely.

Pursuing a strategy mentioned earlier, Duke Power stepped up its efforts to enhance profitability by developing new business both within and beyond the scope of the regulated electric utility business. The goal was to capitalize on the company's existing strengths and experience while at the same time minimizing capital commitments. Through its Management and Technical Services (MATS) section, Duke began to market its expertise in designing, building, and operating power plants. Established in 1982, MATS had already undertaken 44 projects for 22 different clients throughout the country. Through MATS, Duke hoped to stay at the forefront of new technology, and the new unit was selected to perform the design and engineering work for a new coal-burning process and demonstration plant to be operated by TVA in western Kentucky.

The roles of both Mill-Power Supply Company and Crescent Lard and Timber Corporation were expanded. Mill-Power Supply, for example, moved beyond its traditional wholesale electrical supply business and began selling advanced energy management systems and controls to a variety of businesses and industries. It opened its third major distribution center in the Carolinas in 1983 and acquired an electrical equipment distributor in South Carolina. Crescent Land and Timber explored the potential for mineral deposits on its land and embarked on real estate development with the creation of a new business park south of Charlotte.

Duke Power's subsidiaries contributed only $10.4 million to after-tax earnings in 1983, but the company was candid about the fact that it was pursuing expanded unregulated activities with deliberate caution. It warned that the earnings growth from that source was likely to be moderate in the near-term.

Moderate earnings growth or not, Duke Power had finally managed to climb out of its financial hole by the mid-1980s. It had required a wide variety of measures and strategies, but signs of success began to be abundant, and welcome plaudits poured in. "If any utility was equipped to cope with the challenge of the nuclear age," *Forbes* magazine declared, "that utility was North Carolina's Duke Power Company."[47] Equally encouraging were the words from the *Wall Street Journal*: "While other utilities are reeling from cost overruns and operating difficulties at nuclear plants, Duke, the nation's eighth-largest investor-owned electric utility, has compiled a record that is the envy of the industry."[48]

To rest for a while after a ten-year ordeal might have been tempting to some, but not to Duke Power—and especially not to Bill Lee. He agreed with the pronouncement that, "Nothing wilts faster than laurels that have been rested upon." And he proceeded to lead Duke Power through a vigorous regimen as it prepared itself for the new era of competition and the deregulation of the electric utility industry that appeared to be coming in the foreseeable future.

CHAPTER VII

PREPARING FOR A DEREGULATED, COMPETITIVE, AND GLOBAL FUTURE, 1985–1994

The idea of freeing various sections of the American economy from the heavy hand of regulation by the Federal and state governments began to rise to prominence in the 1970s. During the administration of President Jimmy Carter (1977–1981), the airline, trucking, and railroad industries began to be deregulated. Then in the 1980s, during the two terms of President Ronald Reagan, the notion that unfettered competition in a free market would best lead to optimal economic results grew even more attractive to a majority of the American voters and of Congress. This resulted in the deregulation of the telecommunications, financial service, natural gas, and some other industries, and that policy seemed clearly to be the irresistible wave of the future.

Public interest in the deregulation of the electric utility industry became manifest first in California and the Northeast, where electric rates tended to be the highest in the nation. In the South, where electricity was generally cheaper than in many other parts of the country, there was less interest in pushing for deregulation, which was touted by its champions as a certain pathway to cheaper power via competition. North Carolina, for example, would not take its first step toward deregulation until 1997, when the state legislature provided for a special commission to study the matter.

Despite the South's laggardness in this area, Duke Power began preparing itself in the last half of the 1980s for the deregulation and competition that seemed sure to come, sooner or later. Moreover, as a low-cost producer of electricity, Duke approached the new era with great confidence, even with enthusiasm.

Those elements of confidence and enthusiasm came through loud and clear in the company's annual report for 1986, which top management hailed as a "paramount year" in Duke's history. Bill Lee and Doug Booth, chief executive officer and president respectively, offered nine reasons for their claim: 1) earnings for common stock reached a record $4.04 per share (from $3.72 in 1985); 2) Catawba Nuclear Station's Unit 2, the last of seven large nuclear generating units, had been brought into service "ahead of schedule and at lower cost than any other nuclear unit of the same vintage;" 3) following twelve consecutive years of operating the most efficient fossil-fired generation in the nation, Duke employees operated the company's eight coal-burning plants at new record efficiency levels in 1986; 4) Duke's average customer was without electricity only 34 minutes in 1986, a record reliability of 99.9 percent and solid evidence of Duke employees' commitment to quality and reliability; 5) the regulatory commissions in both Carolinas found Catawba Unit 2 to have been prudently built and its capacity needed, which meant that its costs were fairly covered in rates in both jurisdictions; 6) Duke employees made even greater strides in productivity by meeting seven of ten performance goals; 7) nuclear power generated 53 percent of Duke's electricity, saving customers $375 million in fuel costs; 8) in late 1986, Duke benefitted customers by volunteering a 2.3 percent rate decrease, thus passing on savings in income taxes expected from the Tax Reform Act of 1986; and 9) the company's diversified business continued to grow, and on January 1, 1986, the successful engineering services business (MATS) was incorporated as a new subsidiary, Duke Engineering and Services, Inc.[1]

Duke Power intended to minimize the need for new capital by controlling the growth of summer peak demand, by extending the life of older coal-fired plants, and by adding cost-effective peaking capacity (such as pumped storage units and gas-fired turbine units) when needed. With those efforts, Duke believed that its existing baseload capacity was sufficient to serve customers into the new century. The company's hope was that it was setting a course for the future that would position it to shape whatever changes were coming to the electric utility industry, to the benefit of both customers and investors.

Looking to the future, several of Duke Power's top managers expressed their views. "We want to be the best there is in the electric utility business," Bill Lee declared, "and that will be our principal focus as we move into the years ahead." Duke would remain heavily committed to its "core business," which was "providing reliable electricity to customers at reasonable cost." In diversifying beyond that, Duke would look at other opportunities in the light of whether they would enhance shareholder value without negative impact on the company's electric customers.[2]

Doug Booth took a slightly different approach. "We want to emphasize a company culture that is customer-driven," he proposed. That would be essential as competition in the industry intensified. As a low-cost producer of electricity, Duke started with an advantage, and it was "simply going to keep unnecessary costs out of [its] operations." Duke employees had already identified ways to save millions of dollars through the Corporate Goals Programs, and the hope was that they would do even better.[3]

Bill Grigg, formally the executive vice president for finance and administration but more widely known as the chief financial officer, put Duke Power's strategy for the future most succinctly: "We plan to reduce costs as stringently as we can, continue to operate our system as efficiently as we can as we have in the past, increase our kilowatt-hour sales and expand our non-utility income."[4] But, to focus on just one of those measures, could Duke expand kilowatt-hour sales without adding new generating units? The answer was that Duke had launched an aggressive marketing program to increase its share of the residential heating market (where natural gas was a major competitor) by promoting high-efficiency heat pumps. If Duke could increase its share of the residential heating market from the existing 41 percent to 50 percent, that would significantly increase revenue without accelerating growth in the summer peak demand.

Duke was also energetically pushing another off-peak standby that dated from the 1950s, the "dusk-to-dawn" outdoor light, and other types of residential outdoor lighting. Another type of off-peak sale the company had set about increasing was temporary bulk power sales to other utilities.

In pursuit of the last strategy, in 1987 Duke Power negotiated a contract to supply Carolina Power and Light of Raleigh with 400,000 kilowatts for six years beginning in 1992, which would help make more efficient use of Duke's baseload capacity. An even bigger, more long-lasting move was Duke's purchase of the Nantahala Power and Light Company, headquartered in Franklin, North Carolina, in the western part of the state. The addition of that wholly-owned subsidiary opened a market for bulk power sales, mostly off-peak, to supplement Nantahala Power and Light's hydroelectric generating system, which then served about 43,000 mostly residential customers in five mountainous counties.

Duke set about building a transmission line linking it with the newly acquired company. That would assure the customers of a reliable source of power as well as provide Nantahala Power and Light with lower rates than it had been paying to TVA for supplemental power. Moreover, the new transmission line would provide Duke for the first time a transmission link to TVA, thus providing a potential for additional temporary bulk power transactions.[5]

"Duke Puts the Fun Back in Marketing," a headline in the *Duke Power News* proclaimed. More than 400 of the company's marketing people gathered in Charlotte to hear the "whole nine yards" about the new marketing program, the first such across-the-system assembly in twenty years. "In the last 10 [actually closer to 15] years or so," a marketing officer noted, "we've been telling people how not to use our product." Now Duke was "back to the business of competing and selling off-peak electricity."[6]

The splashy program began with a video take-off of the movie "Top Gun." Then on a darkened stage, no doubt with rousing music filling the hall, emerging through a cloud of dry-ice vapor into a bright spotlight was none other than—Bill Lee. Dressed in an aviator's outfit, he gave the crowd a formal salute before delivering a rousing pep talk. (Who said a top-flight CEO should not also have a streak of ham?)[7]

With gas companies, municipal systems, and rural electric cooperatives all vying for customers, Duke Power took on a competitive posture even in the absence of deregulation. The *Annual Report* for 1987 proclaimed: "The Piedmont Carolinas' energy market place of the 1980s and '90s is a prize well worth the battle: a bustling Sun Belt economy flush with affluent, discriminating consumers." The residential market was growing fast as the area experienced a surge in housing construction that reflected both population increases and the maturing of the Baby Boom generation.

Research revealed that the average home buyer in Duke's service area was then 39 years old and increasingly from a two-paycheck family. Many of these Duke Power customers traded up for larger, more modern homes. In Charlotte and Mecklenburg County, one of the fastest growing areas in the Piedmont Carolinas, more than 40,000 new housing units had been started in the previous five years. Likewise, in Durham and Chapel Hill in the Research Triangle area, nearly 16,000 new homes had gone up in the same period.

Since virtually all of the new homes being built had air conditioning, Duke wanted to make every effort to heat the homes in the winter that it cooled in the summer. To accelerate the trend of electric heating, which had begun in the 1950s (with the Gold Medallion Home program), Duke came up with the concept of the Maximum Value Home or, simply, the Max. It combined advanced energy-saving features with a high-efficiency heat pump and electric water heating—the latter available at a half-price rate if used only during off-peak hours and on the weekends.

Residential usage of electricity was by no means the only target of Duke's new marketing strategy. Pointing out that more energy was required to make a ton of paper than to make a ton of steel, Duke Power reported about an energy-saving program that it had worked out in collaboration with its largest one-location customer, Bowater Carolina Company in Catawba, South Carolina. One of the nation's biggest paper mills, Bowater installed a new process to improve paper quality, but, at the same time, the company looked hard for ways to cut energy costs.

Bowater's new process produced large amounts of steam while grinding wood into pulp, and the company's idea was to use the steam to dry paper. After trying one procedure and finding problems with it, Bowater studied the technology offered by Duke Power's heat recovery program and found the savings offered to be compelling.

Bowater then decided to run the steam through a compressor and produce high-pressure steam with two large heat pumps, which would use electricity almost continuously through off-peak periods at night and on weekends as well as during week days. At the same time, Bowater would use less fuel than previously to produce steam for drying. Duke's idea was to push the heat recovery program with other large customers—such as textile, food processing, and chemical manufacturers—that could recover large amounts of energy from steam or hot water.[8]

At the same time that Duke Power returned to aggressive marketing (with an off-peak emphasis), it continued to emphasize an aspect of its corporate life that did not relate directly to the financial bottom line: community service. Carrying on attitudes and practices first fostered by James B. Duke himself, the company exerted itself to be a good citizen, not only in Charlotte but throughout its service area.

In 1952 John Paul Lucas, Jr., then Duke's director of public relations and subsequently the vice president for public affairs, created Duke's well known "bottlecap" logo or seal with the words "Citizenship and Service" emblazoned on it. To attempt to describe the various ways that Duke Power managers and employees tried to live up to and give tangible expression to the company's creed would require at least a long article, but some selected examples might suggest how the company tried vigorously to promote and exemplify community service.

The advocacy of community service by Duke's top managers was not a matter of "Do as I say, not as I do." For example, when John Paul Lucas, Jr., retired in 1975, after thirty-five years with the company, *Duke Power News* noted that some of his civic responsibilities had been his service as vice-chairman of the North Carolina Board of Higher Education; trustee of Johnson C. Smith University and the University of North Carolina-Charlotte; director, United State Chamber of Commerce; district governor, Rotary International; district lay leader, Methodist Church; president of the Charlotte Executives Club; and vice president of the North Carolina Literary and Historical Association. (He also held various professional offices.)[9]

While a comparable list could be submitted for virtually all of Duke Power's top managers, perhaps an indication of some of the civic activities of Duke's rank-and-file employees is equally telling. Carl Horn gave a public address in 1973 in which he summarized the history of the power company and had this to say about Duke Power employees: the company's "most valuable asset is one that does not appear on its balance sheet: its people." Their "morale, abilities, and dedication" never ceased to amaze, Horn declared, and over 3,000 of them had been with the company fifteen years or longer. In addition to serving the company, however, Horn suggested that they also enriched their communities: 1,414 of them were Sunday school teachers; 1,444 were church officers; 751 were Scoutmasters and advisers; 690 were Little League baseball coaches and umpires; 151 were officers in the Parent-Teachers Association; 516 were volunteer firemen; 420 were officers in civic clubs; 115 were volunteer teachers; and many others held various public service positions.[10] While the activities of the employees are themselves interesting, the fact that Horn had the data collected and then publicly reported also says something about both him and Duke Power.

When five men from the electrical construction section at McGuire Nuclear Station used their vacation time to help provide and restore electricity at a church-sponsored hospital in Haiti, the *Duke Power News* carried the story.[11] Soon after, the Greater Carolina Chapter of the American Red Cross cited Duke Power for outstanding participation in the blood-donor program. A substantial number of Duke employees had given several gallons of blood, with the list being topped by a nine-gallon donor.[12]

Carl Horn in 1981 noted that most of the public utility trade periodicals that kept statistics on generating efficiency, construction costs, rates, and other such matters placed Duke Power at the top in both efficiency and economy. "I am convinced," Horn added, "that there is a direct connection between [Duke employees'] extraordinary involvement in civic, religious and educational causes and [their] superior job performances that causes Duke Power Company to be consistently rated 'No. 1.' We are 'No, 1' because of our dedication to the services of the public."[13]

Bill Gregg chaired the United Way campaign for Mecklenburg and Union counties in 1984. When Duke employees became the first employee group in the campaign to contribute more than $1 million, Grigg received a congratulatory letter from President Ronald Reagan. In the entire service area, Duke employees gave $1,757,000 to the United Way campaign in 1984, and their gifts, coupled with the company's corporate contributions, made Duke and its employees the largest single contributor in the two Carolinas, with a total of nearly $2.3 million.[14]

In addition to what individual Duke Power managers and employees contributed to their communities, the company itself strove from its beginning to be a good citizen. Recreational access to and free municipal water supplies from Duke's numerous lakes and reservoirs scattered across the Piedmont have already been mentioned, as have park lands given by the company in both Carolinas.

Duke Power received the National Wildlife Federation's Corporate Award for 1984, and the Federation's spokesman paid the company this tribute: "What sets the Duke Power Company apart in the corporate world is its participation, involvement and leadership in environmental issues." Among its other accomplishments, the Federation declared that Duke had assembled the largest and best in-house environmental staff among the nation's electric utilities. Another aspect that led to Duke's recognition was its outstanding record for the efficient production of power, which resulted in a conservation of natural resources and generated less waste to return to the environment. The Federation also noted that in a highly regulated industry, operating in the absence of external competition, emphasis on excellence and efficiency had to come from within.[15]

In 1982 Duke Power introduced two new community action programs that were perhaps more directly visible than the company's environmental activities. Through its 96 local offices, Duke recruited and trained volunteers from churches, civic clubs and the community at large—including many Duke employees—to weatherize the homes of more than 1,700 low-income families, using materials provided by Duke.

Through a new Community Challenge Heating Fund, Duke contributed $1 to designated community agencies for every $4 they raised to help poor people pay their heating bills. The company committed up to $100,000 for this program initially, and both it and the weatherization program were continued and expanded in subsequent years.[16]

Duke Power introduced a second program in 1985 to raise money for helping low-income families pay winter heating bills. Through the new "Share the Warmth" program, the company included in electric bills a one-time appeal for customer donations. Then customer and company contributions together made more than $652,000 available for home-heating assistance in the service area.

These community-service efforts concerning low-income weatherization and heating earned Duke Power special recognition in 1985: President Ronald Reagan presented Duke's representatives with a commendation at a White House ceremony as part of his citation program for private sector initiatives. The award honored Duke for "exceptional contributions to volunteerism," particularly relating to home heating for low-income people.[17]

While home heating for the needy might seem a natural target for Duke Power, the company also involved itself, in a wide variety of ways, with education. That too, of course, carried on a tradition begun by James B. Duke and his Duke Endowment. Through Duke's matching gift program, for example, its employees contributed nearly $144,000 to 144 colleges and universities in 1983.

Duke began a pre-college development program for minorities in 1986 and through it brought high school students for ten-week summer internships at Duke Power offices throughout the Piedmont Carolinas. In addition to the pay that the students received, Duke set up a savings account for each student where the company matched up to a quarter of the intern's earnings, with a limit of $600. To be eligible for the program, students had to be in the top 10 percent of their class and have an interest in accounting, engineering, science, or liberal arts.[18]

Through a "Power in Education" program begun in 1984, Duke Power tried to marshal its resources to improve primary and secondary education in the Piedmont Carolinas and to encourage other businesses to do the same. In a related effort, Duke Power awarded six four-year college scholarships and fourteen honorary stipends each year to high school seniors in the service area. "The Duke Power Scholastic Excellence Awards" were to go to students showing exceptional scholastic ability and leadership potential.[19]

On a more modest level of education, Duke Power helped some of its own employees to obtain the equivalent of a high school degree by taking classes at their worksites after hours. With some of the classes taught voluntarily by fellow Duke employees, forth-five employees in the central division's construction and maintenance departments obtained equivalency degrees in the summer of 1987.[20]

Blending education and child care, Duke Power, through Bill Lee, played a key role in Charlotte's acquisition of a top-quality daycare facility in 1987. Alerted by employee surveys about the need for more and better daycare options, Lee assembled a small, select group of Charlotte's business leaders in his office and informed them that Duke Power was prepared to commit $20,000 a year for three years. Then he bluntly asked the other top executives how much they were prepared to give to the project. Within a few minutes, more than $300,000 had been pledged.

"I didn't want to have a very long campaign or a very long meeting," Lee later explained. He admitted that his bluntness had annoyed some of the other executives, then added: "But everybody is thrilled about it today. I mean it led to more child-care spaces per buck invested than almost anything else I've heard of."[21] Fortunately for a good number of Duke employees, the new daycare center was located near the large, new Duke Power Service Center in Charlotte's University Research Park.

The company began a new type of civic involvement when it launched an extensive voter-registration drive in 1984, a presidential-election year. The carefully non-partisan "Power in Citizenship" campaign inspired teams of Duke employees to get nearly 20,000 new voters registered, thus accounting for 11 percent of new voters that fall for Duke's North Carolina service area and a record number for one day in South Carolina. In addition to voter registration, Duke employees worked for certain candidates, sent postcards to friends, and helped get out the vote. Bill Lee estimated that the employees had reached at least a half million voters during the 1984 campaign and had let them know how the candidate stood on various issues.

Duke Power won its second top award from the Edison Electric Institute in 1985—an award given annually to an electric utility that made "a distinguished contribution to the development of the electric light and power industry for the convenience of the public and the benefit of all." The award cited Duke not only for excellence in nuclear plant design, construction and operation, but also for the Power in Citizenship campaign.[22]

The Power in Citizenship campaign, while strictly non-partisan, was not, however, just altruistic civic-mindedness in action. Bill Lee made what a colleague described as a "stem-winder" of a speech about the voter-registration and voter-awareness drive to a large group of Duke employees in 1988. He began by declaring his pride in the employees because "at work, in the community, and in the political arena" they knew what it meant to make a difference, which he noted was what the Power in Citizenship program was all about.

Asking his audience to think back to 1984, Lee recalled a political atmosphere in which "electric utilities were the punching bag for candidates fighting it out in hotly contested races." There had been television commercials showing a little old lady begging the meter reader not to send her any more high bills and dramatic shots of candidates for state and local offices smashing electric meters. Lee declared that certain candidates had eyed the utilities as easy targets, and beating up on them seemed "a sure fire way to get votes with no repercussion," for no one expected the utilities to fight back.

According to Lee, those candidates got a surprise because Duke employees did fight back to show that "beating up on utilities" did not result in a free ride to office. "They found out that Duke Power was more than power plants and transmission lines," Lee continued. "Duke Power was 21,000 voters too—active, involved committed voters who were going to tell their side of the story with vigor to their families, friends and neighbors, and even to strangers!"

Since that 1984 election, according to Lee, there had been no candidate in either of the Carolina "who thought beating up on utilities was a free ticket to office." And Duke Power had learned that it had to be involved in politics, as broadly as the law allowed. "Elected officials see us as their constituents now," Lee noted, "just as they do the many other interest groups that make their views heard." Lee believed that anti-utility and anti-business groups in general had always done a better job than Duke Power in staying involved and pushing their agendas. The result was that their persistent efforts gave them a larger voice than their numbers alone ever would have.

Duke Power's involvement, like that of other businesses, had traditionally stopped with cash contributions. But, Lee noted, candidates got hundreds of those and could not "begin to remember where all the checks came from." When a candidate walked through his or her phone bank on election eve, however, and saw it staffed by Duke employees, "*that* the candidate will remember forever."

Lee concluded his "stem-winder" by observing that no one could know what the 1988 campaign would bring. "But we do know that every day in Raleigh and in Columbia and in Washington and in every county seat the leaders we elect make decisions that affect our well-being, and . . . the well-being of our company and our communities. And we know that if we want to help shape those decisions we have to be involved."[23]

Involvement and commitment were not just rhetorical flourishes for Lee. Extending far beyond Duke Power, his involvement took many forms. For six years he chaired the board of trustees of Queens College in Charlotte. The president of the college later recalled that Lee stopped by his (the president's) office almost every Monday morning to discuss things such as allowing men to enroll at what had long been a college for women only. Lee turned down an invitation to be a trustee at Princeton, his alma mater, because, according to the Queens president, Lee "thought the little school down the street needed his help more."[24]

On a more global level, after the disastrous explosion of the Russian nuclear plant at Chernobyl in 1986—a plant, incidentally, that was radically different in many ways from those in North America—Bill Lee took the lead in organizing the World Association of Nuclear Operators, with representatives from twenty-eight nations initially participating. When the new organization held its first meeting in Moscow in 1989, the delegates elected Lee as their first president. Just as the Lee-inspired Institute of Nuclear Plant Operations had striven after the accident at Three Mile Island to intensify and strengthen all training procedures and safety measures of American nuclear plants, the new global organization quickly moved to establish a communications network linking all of the world's nuclear plants and to standardize safety and training procedures.

The importance of those international efforts was underscored by the fact that, while the United States in the late 1980s got only about 20 percent of its electricity from nuclear plants, in France the figure was then 70 percent, in South Korea 53 percent, and in Japan 31 percent. Moreover, China was already planning for hundreds of reactors.

Bill Lee totally shared the belief of Warren Owen, then Duke Power's executive vice president for the power group, that, "The United States can't afford to turn its back on nuclear power." It offered too many economic and environmental benefits while at the same time affording the nation a rare opportunity for energy independence. If the United States wanted to keep up with the rest of the world, Owen maintained, three things were essential: 1) a standardized nuclear plant design; 2) a reform of the licensing procedure into a one-step process; and 3) a change in the political climate and public attitude toward nuclear energy.[25]

Unfortunately for Owen and Lee, those changes were slow in coming. Orders for new nuclear plants stopped even before Three Mile Island in 1979, and after that, as one prominent figure in the industry declared, "No utility executive in the country would consider ordering a nuclear plant today... unless he wanted to be certified or committed."[26]

The situation was undoubtedly a source of massive frustration and disappointment for Bill Lee especially. Nationally and even internationally known as a foremost authority on nuclear-fired generation, he faced a grim reality that he could not alter. Duke Power's record with nuclear plants was, in many respects, the best in the country and had been from the beginning. It had brought its three units at Oconee on line in 1973–1974 at an average cost per kilowatt of $194, well below the then prevailing industry average of $316. Duke managed to stay well below the industry average after that, too: in 1984, when the rest of the industry was finishing plants with a cost of $1,959 per kilowatt, Duke completed McGuire Unit 2 for $918 a kilowatt. "The art of building and managing a large nuclear project," Bill Lee had explained, "is to be self-critical, to find your mistakes early and do something about them."[27]

Catawba Nuclear Station, the last of Duke's nuclear plants and completed in 1986, cost $3.9 billion or $1,560 per kilowatt. It was a two-unit station, but Carolina Power & Light's Shearon Harris plant near Raleigh was completed in 1987, had only one unit, and also cost $3.9 billion but $3,333 per kilowatt.[28]

Even when it came to the painful business of halting work on a nuclear plant already under construction, a widespread occurrence in the 1980s, Duke Power got off more lightly than many of its peers. Duke had expended over $600 million on its canceled Cherokee plant, but a power company in Michigan decided to abandon construction of a plant after spending about $4 billion on it. Another utility in Indiana canceled a half-completed plant after spending some $2.5 billion on it.[29]

That was only one example of the type of information that Bill Lee and some of his managerial associates tried to make sure that key journalists and other media representatives in the Carolina had. In other words, the company realized that it was important to fully brief certain people in addition to Wall Street analysts. A member of the editorial staff of the *Raleigh News and Observer* recalls that in the 1980s Bill Lee scheduled regular meetings in Raleigh with the staff; that they found the discussions informative and lively; and that no other electric utility serving the state then showed a comparable interest in providing the *News and Observer* with such background information.[30]

In October 1984, Bill Lee and others at Duke Power obviously spent a good bit of time conversing with reporters from the *Charlotte Observer*, for the paper printed a couple of unusually informative stories about Duke Power. The shorter one highlighted the accolades that the company was receiving from the *Wall Street Journal* and other business-oriented publications. Since Duke Power common stock had not hit $30 per share since 1969, the fact that it had just closed a week at $29.375 signaled a hopeful future for the company. An analyst at Charlotte's First Union National Bank declared that the 1980s might turn out to be "the decade" for the electric utilities.[31]

A longer feature story on Duke Power in the same issue of the *Charlotte Observer* provided some valuable insights into the company's culture and operations. After describing Duke's do-it-yourself construction policy and the excellent results obtained even with such challenging, complex projects as nuclear plants, the story honed in on the homegrown aspect of Duke Power's achievement. Lee and Warren Owen stressed that the company did most of its recruiting in its service area, which meant that most Duke employees came from and lived in the communities served by the company.

Moreover, of the 32 executives with the rank of vice president or higher, all but 5 had grown up in the Carolinas. (That would change in the 1990s.) Other utilities had been largely unsuccessful in attempting to entice Duke's top managers to leave the company. For example, Bill Grigg, an executive vice president who grew up in a small town about an hour east of Charlotte, had had three offers in the previous eight years to head other utilities but had chosen to stay where he was.

The *Observer*'s story also pointed out that, while Duke Power was not anti-union, it nevertheless benefitted from being located in an area where labor unions were traditionally weak. Of Duke's 21,000 employees, only about 1,800 were union members. Largely unencumbered by restrictive union work rules, the company could be flexible in making job assignments, and an efficient, computerized work-scheduling system seldom left anyone idle.

While Duke Power's wages were certainly competitive for the Southeast, they did tend to be lower than in certain other, high-cost areas of the country. A nuclear project director in Pittsburgh, Pennsylvania, commented about Duke Power: "They hire and fire as they see fit. You can instill much more productivity in people if you say, "Do it our way or you're out."

The reporter went on to note that Duke Power had developed certain methods of encouraging employee productivity that were exportable beyond the Carolinas. Duke fostered internal competition between plants and departments to counter "the tendency to mediocrity" that Bill Lee saw as "inherent in industries that lack market place competition." Employees were assigned specific goals each year, with raises and promotions dependent on meeting those goals. Employees also received stock bonuses that were tied to Duke's achieving certain company-wide goals.

Balancing the good news with certain less happy matters, the *Observer* story then went into a fairly full account of Duke's financial trouble in the 1970s and its labor trouble with the United Mine Workers in Kentucky.

The long article ended by noting that Duke Power was beginning to think about competition in a less-restricted market place. "Why should inefficient utilities be encouraged to build high-cost plants?" Bill Lee asked. He hoped to see electric transmission lines given the same common-carrier status as telephone lines as well as to see utilities allowed to compete for power sales.[32]

Bill Lee also informed the Charlotte journalist that, short of a radical shift in public opinion and the regulatory climate, he (Lee) did not expect any more nuclear plants to be built in the United States until some time in the Twenty-first century—and then only after they had satisfied critics with a record of safety and economy. That was one reason why Duke Power turned to a massive pumped-storage facility as the most economical way to meet the challenge of a peak load that increased each year. The load management program had succeeded in markedly slowing the rate of growth in peak load, but it still grew and had to be covered if shortages and blackouts were to be avoided.

Duke Power was more fortunate than many electric utilities in the eastern half of the United States in that a portion of its service area—and land holdings—lent itself nicely to the pumped-storage technology. It was used earlier at Joccassee Hydroelectric Station in the Keowee-Toxaway complex near Clemson, South Carolina, and close to the North Carolina and Georgia state lines. Bad Creek, however, was to be the largest hydroelectric station on the Duke system, producing over a million kilowatts as compared with Jocassee's 610,000 kilowatts. (The original Catawba plant in 1904–05, it might be recalled, produced 6,600 kilowatts.)

The building of the Bad Creek Station, which went on from 1981 to 1991, involved the largest earth-moving project in the company's history. Many things about the site challenged the engineers and construction crews: steep terrain, silty soils, old landslides and fractured rock. All of these influenced the design and construction, which had to proceed in one of the wettest climates in eastern North America. The average annual precipitation at Bad Creek was 85 inches, and while the project was blessed with some dry spells, there was one five-week period in 1989 where more than 30 inches of rain fell. The rains culminated on July 3, 1989, with a downpour of 8.68 inches. At one point, rising water threatened to flood the powerhouse. It was "a hell of a two weeks," the project manager, Carey York, later noted, but "it was one of those events where people were really working together" and "it helped us grow as a team."[33]

The Bad Creek (or upper) reservoir was created by damming two small streams with two dams and a dike; the main dam, made of earth and covered with rock, stands 360 feet high and measures a half-mile across. At full pond, the reservoir covers 360 acres, is 310 feet deep, and has a maximum drawdown of 160 feet.

Most of the water in the Bad Creek reservoir, however, comes from Lake Jocassee, the lower reservoir. (Lake Jocasee is itself the upper reservoir of Jocassee Hydro Station, for which Lake Keowee is the lower reservoir.) During off-peak hours when the demand for electricity is low, water is pumped from the lower reservoir back into the upper one for storage. Although the pumped-storage process actually consumes more electricity than it generates, the key to its use is the time of day it is operated and the types of other generating sources it replaces. According to Duke Power, pumped storage, when used in a prudent generating scheme, can be one of the most economical forms of electric power generation available.

When power from Bad Creek is needed—say on a quite cold winter morning or a hot summer afternoon—water stored in the Bad Creek reservoir is released into a long underground power tunnel. Then, like water draining from a sink, it rushes down the tunnel and drives a set of huge turbines before emptying into Lake Jocassee. The entire power tunnel is close to ¾ of a mile long, almost 30 feet in diameter, and initially drops more than 850 feet vertically. The powerhouse chamber is 540 feet underground in solid rock and is so vast that a 32-story building, turned on its side, could fit inside the powerhouse.[34]

Bad Creek's dams, reservoir embankments, tunnels and powerhouse are "wired with more instruments than a patient in intensive care." The purpose of all of the "geotechnical instrumentation" is to provide ongoing data about the performance and stability of the structures and to enable engineers to keep a constant check on the structures' performance as compared with the design assumptions. All of the remote instruments are battery powered with solar recharge and networked into a computer.

During construction one would have had difficulty imagining this, but before the Bad Creek reservoir was filled, it was cleared of all potentially floating debris larger than 3 inches. That was done because, in order to minimize "head loss," Bad Creek's intake was not equipped with a trash rack.

Those who worked on the project took pride in the fact that they beat the construction industry's average of 5 percent for "rework," or work that had to be redone for one reason or another. At Bad Creek rework was reduced to 2 percent because, according to a Duke Power spokesperson, quality was designed and built into each part of the plant. The employees from the quality assurance department worked closely with the craft and engineering employees from all departments, and the craft supervisors participated in the same training as the quality assurance inspectors.

Bad Creek's safety record was another source of pride: there were no on-the-job fatalities during the construction of the massive project. Moreover, from 1988 through 1991 the employees on the project worked more than 3 million hours without a lost-time accident. "Safety came first," the project manager explained. "We realized that if our people knew that we put their safety as the first priority, then good things would follow. We ranked safety first, then quality, then cost and then the schedule."[35]

An aspect of the schedule became a final matter of pride for the employees involved. Several years after construction began in 1981, the project had fallen several months behind schedule. By mid-1988, therefore, the schedule was revised to provide for the completion of units 1 and 2 by April 1, 1992, and of units 3 and 4 by January 1, 1993.

Someone, however, became inspired to come up with a catchy rallying cry for the Bad Creek labor force: "All done in '91." Emphasizing team work, the men and women at Bad Creek adopted the following mission statement: "To provide Duke Power Company with a reliable power plant at a cost savings of $200 million by having all four units operable in 1991."[36] And they did it.

Bert Tompkins, a mechanical engineering supervisor who spent six years at Bad Creek, noted that when he first went there the project had a bad reputation. "Nobody wanted to go there," he explained, "and one reason was all the stories about snakes up in the mountains." Toward the end of the project, however, "everybody wanted to be there," for the "teamwork at Bad Creek was more than just a story or a headline." He thought it "amazing to see a bunch of people work so hard and have such a good time."[37]

Those involved at Bad Creek may not have fully realized it, but the project was something of a climax for Duke Power's ventures in hydropower as well as a "last hurrah" for the company's proud tradition of do-it-yourself construction, at least in the Twentieth century. The company's next generating plant was destined to be a gas-powered combustion turbine facility, and, rather than being built by Duke employees, it would be put out for competitive bidding.

Costing about $1 billion, Bad Creek compared favorably with other pumped-storage plants. Bad Creek's cost was an estimated $932 per installed kilowatt, but a survey of several pumped-storage projects built by other utilities in roughly the same time frame revealed costs per installed kilowatt ranging from $964 to $1,626 in 1991 dollars.[38]

Duke Power anticipated a reduction in work force upon the completion of Bad Creek. Accordingly, the company utilized a larger-than-normal number of temporary-contract workers on the project.

In another sensitive area, that of the environment, Duke Power took some imaginative measures at Bad Creek. When choosing a site for the station, Duke Power scientists conducted extensive field studies and, among other things, they found such rare plant species as the wildflower Oconee Bells. Either working around them or relocating them before construction began, the Duke employees were praised by conservationists for their efforts.

Faced with finding an environmentally sound way of disposing of the huge quantities of rock excavated from Bad Creek's powerhouse chamber, Duke engineers found that they could use the material to solve another problem. Lake Jocassee was renowned for its record-breaking trout, which lived in a thin band of cold, oxygen-rich water. If it should be disrupted and infused with warm surface water, that would affect the ability of the fish to feed and grow. Duke's engineers and botanists determined that they could help avoid that by using the excavated rock to build an underwater dam, or weir, out from Bad Creek's discharge portal. The weir would break the force of the discharge and thereby preserve Lake Jocassee's delicate water balance.

For these and other such environmentally-inspired measures, Duke Power in 1987 received a Merit Award from the Soil Conservation Society of America. Likewise, South Carolina's Land Resources Commission and Oconee County's Soil and Conservation District saluted Duke Power's conservation efforts at Bad Creek Station.

Unlike the much larger Lakes Jocassee and Keowee, Bad Creek's upper reservoir was not open to the public because rapidly fluctuating water levels presented a potential safety hazard. Public access areas on both Lake Jocassee and Lake Keowee, however, made them available for boating, swimming, and tournament-caliber fishing. And Duke Power collaborated with the South Carolina Department of Parks, Recreation and Tourism to develop the Devils Fork State park on the lower end of Lake Jocassee.

Duke Power also designed and built a 43-mile segment of the Foothills Trail; the segment, built according to Appalachian Trail standards, linked together two previously existing trails. For that, as well as other recreational opportunities provided by the company, Duke received a Recreation Achievement Award from the South Carolina Outdoor Parks and Recreation Exchange Council.[39]

Despite all the environmental care that Duke Power had taken and the awards the project had received, Bad Creek and other pumped-storage facilities like it were anathema to many environmentalists. This was so for several reasons. First, a massive hole had been gouged in the land—at vast expense—in a scenic, pristine area. Then, since the technology of pumped storage involved the use of more electric power than the plant actually generated, that meant that those coal-fired and nuclear plants that then supplied Duke Power's baseload were either pumping more pollutants into the atmosphere or adding to the mounting supplies of nuclear waste, the long-range storage of which remained a thorny problem that the nation still faced. Either way, countless environmentalists were distinctly unhappy. For that reason, among others, plans for additional pumped-storage facilities were not to be on Duke's drawing boards as the century neared its end.

What was on the drawing boards were the strategic plans of Bill Lee and the other top managers to prepare Duke Power for the new era in the electric utility industry that they perceived, quite correctly, to be emerging. Change does not come easily and painlessly to many human institutions, of course, and a large, venerable utility like Duke Power was certainly no exception. Yet change, sometimes wrenching and achieved as rapidly as possible, was precisely what Bill Lee and his colleagues were determined to have.

Already in competition with the natural gas industry and the cooperatives, Duke Power, according to Lee, was soon destined to be slugging it out with other utilities. Rate hikes, therefore, were to be avoided as long as and as much as possible, and costs had to be not only held down but reduced wherever feasible.

So much good news came Duke Power's way in the late 1980s that it heightened the shock of the inevitable downsizing that lay ahead. As one example of the pleasant news, *Institutional Investor* magazine, an important trade journal in the financial world, named Duke's Bill Grigg as one of the ten best financial officers in the United States. The selection was based on recommendations from analysts, investment bankers, and other lenders, and those chosen had to have "a substantial record of accomplishments" in jobs that were "especially challenging." The magazine gave Grigg credit for "managing to underwrite Duke's construction program while slowly but systematically restoring Duke Power's double-A rating" when times were tough in the 1970s. One banker was quoted as saying, "Grigg is a cut or two above anyone else in the [power] business." Another reported that "while Grigg's courtly style makes him a pleasure to work with," he was also respected for "his underlying toughness."[40]

While Grigg received such warm plaudits, so did Bill Lee. Named by the *Wall Street Transcript* in 1987 as the best chief executive of electric utilities east of the Mississippi (for the third straight year), Lee gave an interview to a reporter for *The Energy Daily*. The story began with a quotation from a Wall Street analyst who declared that Lee ran "the tightest ship in the business." Then Lee explained that Duke employees were "very excited by goal setting and working toward goals." He thought the company had developed "the right incentive goals program" that applied to all employees at whatever level. Duke "sweeten[ed] the match on the stock purchase program."

Citing Duke Power's long-running record for having the most efficient coal-burning plants in the nation, Lee gave this insight into one of the explanations for that record: "We have a guy going off shift at 11:00 p.m. and calling his buddy at a plant 110 miles away, and he says, 'How did you do on your shift? I bet I beat you.'" When one gets that sort of thing going, Lee continued, "you know something good is happening." The incentive was a few more shares of company stock, but Lee believed that the stock was just a symbol and that the recognition and sense of accomplishment were what really mattered.

"And that spirit is throughout," Lee boasted. Duke had just brought another nuclear plant (Catawba) on line "at the lowest cost per kilowatt... in the country." Moreover, the plant came in "nine months early and $200 million under budget." Lee noted that when Duke designed a power plant from scratch, the company brought in the operators and assigned them to the design department. "And they can tell a theoretical engineer like me," Lee continued, "'Look, you've got to turn that valve around so I can reach it from the platform.'" Practical operating input while the plant was being conceived and designed helped "the availability of the plant later, the maintainability, and it helped the operators achieve thermal efficiency."

As a result of the accumulated experience and expertise of Duke's engineers and design people, Lee pointed out that there were already some 40 other clients for their services. Duke wanted to be careful, however, "not to oversell the seats on the airplane." That is, it did not want to take on a job where it had to go out and hire new people to do the work. "We want to make sure," Lee maintained, "that they [who seek Duke's services] get exactly our track record, our expertise."[41]

Explaining the selection of Bill Lee as the best chief executive of electric utilities in the eastern half of the nation, the *Wall Street Transcript* noted that the successful completion of Duke's nuclear construction and "smooth regulatory sailing" contributed to the favorable forecast for Duke Power, as did a strong cash flow. "Analysts following the utilities industry from virtually every quadrant point to Duke as a winner," the business journal asserted. Moreover, Lee received much of the credit for the company's "outstanding day-to-day operations skills," as exemplified by Duke's long-standing supremacy in the efficiency of its coal-fired plants.

Given the reasonably favorable growth prospects of Duke Power's service region and its close attention to costs, the company got high marks from those polled not only for past performance but also for future prospects. Finally, the *Wall Street Transcript* pointed out that Bill Lee had gained recognition "as a chairman with an entrepreneurial approach to a changing industry."[42]

Doug Booth, Duke's president and chief operating officer, reported that one way in which the company sought to gain and keep a competitive edge was through "quality circles." Duke's first began operating in 1981, and by 1987 there were 165 active circles representing 13 departments across the company. Each circle consisted of from 4 to 10 volunteers, either from the same work group or related groups, and they met approximately an hour a week to identify, analyze, and solve work-related problems. Having visited a number of the circles, Booth said that he had been impressed with the spirit of teamwork and pride each circle took in its role within the larger organization. Other companies in the Carolinas were turning to Duke Power for help in establishing their own quality circles.[43]

Against this type of up-beat and encouraging background, it is no wonder that when downsizing—so dreaded by workers and beloved by Wall Street—came to Duke Power in 1988, the shock was painful and long remembered. Though discharged employees quite understandably had trouble accepting the argument, Duke's leaders had solid reasons for the action.

As part of management's strategic plan and on the heels of a series of highly unpopular rate hikes in the 1970s and early 1980s to pay for nuclear units (and keep up with inflation), Duke Power had promised its customers that, barring unforeseen circumstances, there would be no rate increases before 1992, when Bad Creek was scheduled to come on line. Yet in 1988 it became apparent that drastic steps were needed if Duke were going to keep its word to its customers and remain attractive to investors.

Accordingly, in mid-1988 the company initiated a work-activity review and asked each department to make a careful study of all activities; quantify the least important 10 percent of those activities; explain what the consequences would be if they were eliminated; and relate the results to headcount reductions. As Ken Clark, then the vice president for corporate communications, admitted, the process was "no fun." Like most utilities, however, Duke Power, according to Clark, had acquired a reputation, among employees as well as outsiders, that a person, once hired, had to work hard to get fired. As a result of the work-activity review, however, in November, 1988, Duke laid off or gave early retirement to more than 6 percent of its 20,000-person work force. That translated to 1,200 jobs. "As a company," Clark declared, "we're now learning how to be leaner and meaner and more competitive...." He believed that the agonizing process had been not only necessary but also a change of direction that was essential if the company wanted to survive in the competitive era then unfolding.[44]

Bill Lee, at whose desk "the buck stopped," had to make the final decision to cut the work force, and he declared in a published letter to all employees: "I have personally never been through a more traumatic experience in my life." He went on to explain that Duke Power did not make the job cuts because the company was in a financial crisis, but it made them to prevent one in the future. "The layoff was driven by the absolute need to hold our prices steady," Lee continued, "and the only way to do that was to further cut our costs." And why was it so essential to hold prices steady? The answer was simple: to meet increasing competition on several fronts. What Lee did not say was that the company's management was acting on the assumption, which proved quite correct, that competition in the power business would soon grow even more intense.[45]

The workforce reduction of late 1988 was only one of a series of measures that Duke Power took to become more competitive in preparation for the future. Considering the growing probability of deregulation and increased competition, Bill Lee pointed out in 1986 that there were technical, political, and economic questions that were still unanswered. "But I think you will see Duke anticipating and shaping change," he added, "and not being dragged along by it." From "an attitude point of view," Lee concluded, "we are willing to compete, head-to-head, under any circumstances, with anybody as long as it's fair competition."[46]

Even as a regulated monopoly, Duke Power had never been indifferent towards its customers or about service to them. In 1990, however, the company began to use a new slogan or mantra: it now wanted to be the "Company of Choice"—the supplier of choice by its customers, the employer of choice by its employees and the communities they served, and the investment of choice by its owners. It was a tall order, but to become the Company of Choice, Duke identified in 1991 seven areas of strategic focus: customer satisfaction, financial management, nuclear excellence, team excellence, environmental leadership, expanded business opportunities, and excellence management philosophy.[47]

Putting its money where its mouth was, as the old saying goes, Duke Power announced in 1990 that it was building a new state-of-the-art Customer Service Center. Employing about 400 specially trained people, the innovative facility would handle inquiries from all over the service area 24 hours a day, 7 days a week. Duke suggested that no other utility in the nation offered as high a level of service to call-in customers.

Having special needs and problems, Duke's largest industrial customers received tailor-made attention and programs. The top 20 industrial customers accounted for sales of more than 10.2 billion kilowatt-hours and $396 million in revenue per year. The number one industrial customer, Hoechst Celanese, produced chemicals, fibers for textiles, plastics, pharmaceuticals, and other products in nine major plants served by Duke.

The second-ranking industrial customer, Bowater Carolina Corporation, produced pulp, newsprint, and coated paper. Energy was one of Bowater's top four expenses, and electricity accounted for between 70 and 80 percent of its energy dollars. Textile mills, as one would expect, occupied a significant number of the top-twenty slots, and coming in towards the bottom of the list were both Duke University and the University of North Carolina at Chapel Hill.[48]

All of the emphasis on customer service and customer satisfaction obviously paid off, for beginning in 1993 and for several years thereafter surveys conducted by an independent research firm revealed that 88 percent of Duke Power's customers believed that the company operated in their best interest. The same percentage revealed that they were either very or somewhat satisfied with Duke, and around 80 percent found that Duke's electric rates, as compared with other products and services, were reasonable.[49]

A comparison of customer satisfaction done by a national research group revealed that 78 percent of Duke's customers were "very satisfied overall" and 97 percent gave the company an overall reliability rating while the national figures were 56 percent and 72.6 percent respectively.[50] All this good news for Duke Power about its customer relations was capped by the results of a survey conducted by the business school at the University of Michigan and the American Society of Quality Control: Duke Power ranked as the nation's top electric utility in customer satisfaction.[51]

While the news about Duke's customers was understandably encouraging, survey results concerning the thinking and attitudes of Duke employees were much more of a mixed bag. The company conducted such surveys on a regular basis, and the monthly news journal for the employees reported the results in a candid fashion. No doubt with the memory of the 1988 layoffs still much in mind and additional changes still underway, a survey in 1991 revealed that there had been a decline in employee morale and that there were mounting worries about job security.[52]

Inviting employees to submit questions for management, the *Duke Power Journal* selected this one for Bill Lee: in light of the survey results, what were company leaders doing to restore trust in management? Lee, admitting that the 1988 layoffs had been "a real jolt to our culture," suggested that the easiest and quickest way to remedy the problem would be to say that the downsizing was a mistake and would never happen again. "A more difficult thing," Lee continued, "is to convince you that in a competitive world your only job security is to serve the customer in a world class company that can out perform anybody—as opposed to an artificial promise that a leader might make. Ultimately, trust will be earned by performance."[53]

By 1994, when Duke employees owned over 17 million shares of stock in the company, the employees continued to "signal a wary distrust of company management...," as the *Duke Power Journal* declared. The survey had good news about employees' ratings of supervisors, pay and benefits, safety and local quality measures, and all of those things got better marks than in the 1992 survey. But as for the still-lingering wariness about management, the vice president for organization services, Chris Rolfe, confessed: "... When I look at what employees have been through the past few years, it's not hard to see that, as a group, we've had a big, big serving of change—maybe almost too much to swallow for some of us."[54]

One employee in customer operations posed a thoughtful question for management: "If we [at Duke Power] are as good as we've always heard we are, why do we need to change so much of what we do?" The newly-named senior vice president in customer operations, Jim Hicks, replied: "Without a doubt, we are an industry leader. But the industry we are a part of is going through some huge changes, and we've simply got to change if we expect to improve on [or even hold?] our position...." He went on to explain that the pressure for change was coming from many sources: regulators were easing restrictions on where large customers could buy electricity; evolving technologies were giving customers alternatives to buying power from a utility; and customers' businesses were changing and causing them to rethink their expectations of Duke Power.[55]

While all that Hicks said was quite true, the fact remained that Duke Power had an ongoing problem about employee morale. In 1996, the editor of the *Duke Power Journal* reprinted some brief, random quotations from the employee surveys and then included what he described as a "composite" quotation: "My future at Duke is out of my control. I don't know what to do." The vice president for organization effectiveness, Chris Rolfe, could only respond: "...It may be a rationalization, but companies all over the country are hearing the same messages from their employees—our survey results track right along with national norms and benchmarks....This tells me we're wrestling with the exact things that people all over America are wrestling with."[56]

What Chris Rolfe said had been asserted a bit earlier by a nationally prominent consulting firm that Duke Power engaged to conduct a scientific study of its corporate reputation. "Reputations and opinions may be seen as 'soft' issues," a managing partner in the consulting firm noted, "but in tough times, they can often be the tie-breakers." At any rate, the study revealed some happy news for Duke Power when the managing partner declared, "This company is extraordinary... [and] certainly has a powerful reputation in this area." In scores of "corporate equity," which was a calculation of the link between a recognizable company name and favorability, Duke Power kept company with United Parcel Service, Microsoft, and General Electric—and scored higher than either Motorola or Disney.

There was a caution light, however, and it related to employees. According to the study, they were "less enthusiastic about recommending the company as a good place to work and less willing to give the company high scores on the favorability scale." The managing partner had only this limited consolation to offer Duke's management: "It's not just Duke. Corporate America has not exactly done a good job of fostering corporate loyalty."[57]

There was, then, a price to pay in employee morale and company loyalty as a result of all the changes that Bill Lee and his fellow managers believed they had to make, if the company was not only to be competitive but also to come out ultimately as a winner. That all of the predictions about deregulation and increased competition were absolutely correct became dramatically clear in 1992: the Energy Policy Act of that year, in the words of one historian, "greased the skids for greater competition within the electric supply industry." The law made it easier for newcomers (i.e., independent power producers) to enter the power generation business by removing restrictions that had prevailed since the New Deal of the 1930s. Moreover, the new law also opened up the use of a utility's transmission lines by competing power producers, thereby helping them reach new customers. The Energy Policy Act did not permit the Federal Energy Regulatory Commission to order retail "wheeling" (that is, the transmission of electricity directly to a customer by a producer other than the customer's local utility). But the act did not prohibit the states from doing just that—as a number of states soon began planning to do. The newcomers or independent power producers, many of them using the newest form of power generation—gas-fired combustion turbines—would soon be scrambling to try to take business from existing utilities.[58]

Concerning the Energy Policy Act of 1992, which Duke Power had strongly supported, Bill Lee noted that it would quickly open the door to new business opportunities for Duke. At the same time, however, it presented a new challenge to the company to be a cost-effective producer in order to be the supplier of choice for customers. Steve C. Griffith, Jr., Duke Power's general counsel and executive vice president, considered the Energy Policy Act the most significant piece of legislation affecting the industry that Congress had passed in the last half century. But he also felt confident that Duke's strategic plan and quality process were sufficient assets to keep the company ahead of the challenges.[59]

Some prominent industry analysts were already arguing that there were far too many electric utilities in the nation and predicting that there would be a sharp reduction in their number before the new century arrived. An official of the World Bank, who directed its energy and industry department, declared: "I doubt very much the U.S. utility industry as we know it today will survive the decade." He expected that the winds of change, already blowing through the "hide-bound" industry, were about to become a howling gale.[60]

That, of course, was what Bill Lee and his associates had foreseen and been preaching about since at least the mid-1980s. To bring the message home with a particular and painful example, the *Duke Power Journal* publicized something that Duke Power was not accustomed to: it had lost an important potential customer. Parkdale Mills, one of Duke's top 20 industrial customers, would not build its new state-of-the-art textile mill in North Carolina after all. There were a number of factors involved, but an important one was that it got lower electric rates from Appalachian Power for the site chosen for the new mill in Virginia. Appalachian's industrial rate was 3.76 cents per kilowatt-hour, while Duke charged an average of 4.29 cents. (Later in the year, which was 1994, the North Carolina Utilities Commission approved a new economic development rate, which would have strengthened Duke's bid for the Parkdale plant.)[61]

Duke Power, unaccustomed to losing out in a situation such as that with the Parkdale plant, had long enjoyed a different scenario. Bill Lee pointed out, for example, that when a German automotive giant proposed to build a large new BMW plant in the United States, there had been an intense competition among various states that hoped to get the plant. Lee explained that energy costs had been a prime consideration for the BMW people, and Duke Power's rates and service helped South Carolina land the prize.[62]

Bill Lee would have preferred, of course, to tell BMW-like tales exclusively, but his hopes for Duke Power's future in a fast-changing industry would not allow such self-indulgent complacency. With first his national and then his global leadership in the nuclear-energy field, Bill Lee, who had joined Duke Power in 1955, earned so many awards and gold medals that to list them would be tedious. At the time of his retirement in 1994, he confessed in an interview that, while he had developed the idea of the company's mammoth Keowee-Toxaway complex in the 1960s, he had a young engineer's carefree insousiance about where the money would come from. Not until the financial crisis of 1974, he explained, did he learn "a big lesson—that money doesn't grow on trees." Up until that time, what old-timers told him about the great depression of the 1930s "went in one ear and out the other." He learned well in 1974, however, that "we were going to have to build this company to be financially strong."

As for retirement, Lee planned to keep an office in the headquarters building and especially to try to continue helping several of Duke's fast-growing subsidaries, for he had contacts with their potential customers. For years, however, he said he had enjoyed only one or two free nights a week, and he hoped to have more time for his family. "I'm a 'type A' personality, you know," Lee concluded, "and I've got to have plenty of activities, but I want more time for fishing, hunting, gardening, reading." Unfortunately, Lee did not get much time for his family or outside interests, for he died suddenly of a heart attack at age 67 on July 10, 1996.

Several month's prior to Lee's retirement, the board of directors had announced that he would be succeeded as chief executive officer and chairman by Bill Grigg. And it would fall to Grigg to lead Duke Power in making what was surely the most momentous change in the company's history.

CHAPTER VIII

THE CLIMAX OF CHANGE, 1994–1997

The company that Bill Grigg began to lead in 1994 was vastly different from the Duke Power he had gone to work for in 1963. In fact, the company had undergone so many changes in the previous decade that it was quite different from what it had been even in 1984. The biggest change of all, however, was yet to come.

Grigg's style of leadership was different from but no less effective than that of Bill Lee. In an interview at the time Grigg became chief executive officer, he noted that Bill Lee's main challenge had been to complete a massive construction program, to get all those new plants into the rate base while handling the politics related to that, and to see that Duke operated its system with the highest possible efficiency. Grigg believed that Lee and those who assisted him had met those challenges well. "Our focus going forward must be to offer better service at a lower cost than any competitor," Grigg declared.

Duke did not want its industrial customers building on-site generation, he continued, unless Duke built it. And the company did not want any large customer, who might have a choice of suppliers in the market that was emerging, to go someplace else for power. Since Duke's industrial rates were among the lowest in the Southeast, there did not seem to be any immediate threat, but if independent power producers and on-site generators continued to lower their costs, Duke could become vulnerable. He had noticed with grim interest that a Midwestern utility had lost four of its largest industrial customers, and, as a consequence, had been compelled to write off more than $1 billion worth of plant, cut its dividend by half, and lay off employees.

Grigg specified three immediate objectives for Duke Power: 1) "Get totally focused on the customer—every customer;" 2) look for new services and products, related to Duke's core business, to offer to customers; and 3) stay ahead of technological change, which came regularly and rapidly in the whole energy field.

When asked about his style of leadership, Grigg replied, "I believe the leader of a team should be the first one in to work in the morning and the last one out at night." He thought that a great weakness in American business leaders was that many of them lost touch with reality. The temptation was to say, "I've arrived." Then came the corporate jet, the limousine, and all such perks. "I'm very uncomfortable with that," Grigg asserted, and "if I get the limo syndrome, I want somebody to say 'Time out!'"

On another matter, many Duke employees were concerned that Duke Power common stock, which had been up around 45 late in 1993, had fallen to around 35 by April, 1994. Could Grigg explain that? He began by noting that Duke had had a total average return (dividend plus increase in value of the stock) of more than 20 percent over the previous decade. Interest rates had risen, however, sending down not only utility stocks but also other stocks which were sensitive to interest rates, such as those of banks and other financial institutions. When interest rates climbed, investors tended to go for higher-yielding investments. One should not worry about short term changes in stock prices, Grigg suggested, but keep focused on the fundamentals of Duke Power's business.

Since Duke Power was in a maturing industry, Grigg continued, the prospect for growth was only about 2 percent a year. That meant that Duke had to work harder and smarter to keep growth in earnings above inflation—and in a competitive market it could not do that by raising rates.

Grigg confessed that when he joined Duke Power as an assistant general counsel in 1963 he had seen it as an opportunity to "practice law to the hilt," without worrying about how he spent the client's money. (Shades of young Bill Lee!) Then when Carl Horn asked Grigg to become vice president for finance in 1970, Duke soon had to start in on stormy rate cases where the company desperately needed increases despite the public's howls. "All hell broke loose," Grigg declared, and it "was not at all what I bargained for when I came here to practice law."

When asked if he had known any "defining experiences," Grigg mentioned three. First, the third shift at a cotton textile mill had taught him as a young man to appreciate hard work, but more importantly to appreciate the people who do it—"the same folks who operate our power plants, who hang the line and keep the lights on, and serve the customer." They were the people, he added, who also "fight the wars, pay their taxes and work to do what's right."

A second great influence, Grigg noted, had been the two years he put in with the United States Marine Corps (after graduating from Duke University in 1954 and before entering Duke Law School). "I learned that loyalty, excitement and performance don't come from fat paychecks and fringe benefits," Grigg explained, "but from confronting and overcoming challenges as part of a unit.... You couldn't let your unit down."

Finally, Grigg believed that his position as editor of the *Duke Law Review* had been important to him: "That taught me that you can't be sloppy in what you do. Every sentence has to be documented, orderly and persuasive."[1]

Influenced though Grigg had been by an early job in a textile mill, a stint in the Marine Corps, and the top student office in Duke University's law school, his more than thirty years with Duke Power had undoubtedly been the most significant fact about his professional life. Beginning in the "glory years" of the 1960s, he had then played a crucial role alongside Carl Horn and others in navigating the company through the financial perils of the 1970s and into the sunnier skies of the 1980s.

One of Grigg's early moves as chief executive officer was to reorganize, with the backing of the board of directors, top management. The core electric business was growing slowly, and even if Duke Power were the best utility it could be, Grigg explained, "that key market is projected not to grow fast enough to get Duke Power where we want to be." To get the core electric business to the highest possible level of performance, however, its leadership was being consolidated under Richard B. (Rick) Priory, the newly named president and chief operating officer. A native of New Jersey, he graduated in engineering from West Virginia Institute of Technology and then obtained a master's degree in civil engineering from Princeton University. After working briefly for Union Carbide, Priory taught engineering for three years at the University of North Carolina, Charlotte, before joining Duke Power in 1976. He was destined eventually to become Bill Grigg's successor.

With markets and the regulatory environment changing so rapidly, Grigg continued, there was a need for an executive officer who would be "above the fray of daily Duke Power business." Don Denton, as chief planning officer and senior vice president, had been given that responsibility.

The leadership of the company's major support functions—ranging from corporate facilities and human resources to corporate communications—was consolidated under Dr. Ruth G. Shaw, new senior vice president of corporate resources and chief administrative officer. President of Central Piedmont Community College before joining Duke Power in 1992, Shaw first served as vice president for corporate communications.

Finally, since Duke Power was counting heavily on its non-regulated ventures to contribute increasingly to the bottom line (and therefore to shareholders' value), Grigg and the board selected William A. (Bill) Coley to head the new Associated Enterprises Group (AEG). It included Duke Water Operations, Duke Merchandising, Duke Energy Corporation, Duke Engineering & Services, Duke/Fluor-Daniel, DukeNet Inc., Crescent Resources Inc., and Nantahala Power & Light.

Some of these subsidiaries–such as Crescent Resources and Duke Engineering & Services—had been in existence for a number of years; others were relatively new. All were meant to be profit makers as well as, in certain cases, to provide employment for Duke personnel who were no longer needed in the core business. Moreover, it was some of these subsidiaries that increasingly gave Duke Power a global scope.

Duke Energy Corporation for example, had been created in 1988 to plan and carry out the development and financing of new power projects outside of Duke Power's traditional service area. Focusing initially on the building of coal-fueled generating facilities in the eastern United States, Duke Energy led a consortium in 1992 that purchased the majority interest in the Guemes Power Station in Argentina. (In 1997, when Duke Power merged with PanEnergy and the enlarged company became Duke Energy, the name of the older subsidiary was changed.) At the same time, Duke Engineering and Services led the team that operated the Argentina plant.

Duke Engineering & Services, Inc., formed in 1987, focused on nuclear, hydro, and transmission and distribution services. It was an outgrowth of the company's earlier Management and Technical Services division, and one of its initial projects was to design a cogeneration (combined heat and power) facility to supply space and water heating for the United States Army's Fort Drum in upstate New York. Excess electricity from the plant was to be sold to a near-by utility company.[2]

By 1991 Duke Engineering & Services worked on 147 projects for 90 clients and had doubled its workforce. The subsidiary contracted with the United States Department of Energy, for example, to assist with the nation's civilian radioactive waste management program. In 1993 Duke Engineering & Services opened an office in Melbourne, Australia, and provided consulting services to the state government's Electricity Commission. Duke Energy also began looking for additional opportunities in electric generation in Australia and other Pacific Rim countries.

While both Duke Engineering & Services and Duke Energy expanded their global reach, DukeNet Communications, formed in 1994, stayed closer to home. An out growth of the expertise Duke Power had developed in building its own fiber-optic communications network, DukeNet's initial project was to build a fiber-optic network linking homes, offices, and retail businesses in a development in Greensboro, North Carolina. Late in 1994 DukeNet joined with a subsidiary of BellSouth, C.P. & L., and others to bid for a personal communications service license for the Charlotte Major Trading Area, the 6th largest in the country.

Crescent Resources, one of Duke Power's older and most profitable subsidiaries, made an especially strong showing in the 1990s. Building on the continuing success of its golf-oriented residential community on Lake Norman, Crescent proceeded to develop other major residential communities, such as Ballantyne, located south of Charlotte, and Sugarloaf north of Atlanta.

After Crescent developed a 600,000-square-foot distribution center for the T. J. Maxx Company in 1992, Black and Decker selected the Duke subsidiary to build an even larger distribution center near Charlotte, one designed to serve all of the company's customers east of the Rocky Mountains. Crescent's office buildings in various Southeastern cities were also prime commercial properties.

In 1995 Crescent achieved an impressive $35.5 million in net income, a sum that was $13 million over budget and more than double the level reached just two years earlier. *The Business Journal* recognized Crescent as the Number One commercial developer in the Charlotte area, whereas five years earlier it had not been ranked.

At The Summit, a Crescent community on Lake Keowee, buyers came from eleven states and one foreign country to purchase lots, and first-day sales topped $4 million. An even more impressive result came when Crescent put 53 lots on Lake Norman on the market: at the end of the first day, 44 lots sold for more $7.4 million, or an average of 10 percent above the listing price. One lot drew offers from 12 potential buyers. With options in late 1996 to purchase land near Hilton Head, South Carolina, plans for a large office park near Atlanta, and an office building near Nashville, Tennessee, being developed, Crescent was clearly booming.

With real estate development throughout the Southeast proving so successful for Crescent, perhaps it was not surprising that it began systematically scaling back on the forest operations that had been its original purpose. Late in 1996 Crescent announced that it was making available for purchase by conservation agencies approximately 42,000 acres of land owned by Duke Power along the Blue Ridge Escarpment. Hailed by some conservationists as "the most important resources conservation initiative" of the last half of the century, the land included spectacular gorges on several rivers and streams, with scenic value that was said to rival that of Yosemite and other premier natural areas in the United States. Duke offered the land at its appraised value, and for each two acres purchased, Duke agreed to donate an additional one.

While Duke retained some of the land in the Jocassee area for possible future options in power generation, the South Carolina Department of Natural Resources made the first purchase of a 1,000-acre tract near Lake Jocassee, and eventually the entire 42,000 acres were transferred. A senior scientist in Duke's environmental division pointed out that broad-based partnerships between Crescent and Duke on the one hand and resource agencies and environmental groups on the other had been operating in the Keowee-Toxaway area for thirty years or more. Those efforts had resulted in the restoration of Peregrine falcons, deer, and turkey populations; the enhancement of trout habitats; research on black bears; and the location and protection of numerous rare plant species. At the same time, public access and watershed protection had been provided.

"Crescent and Duke can be proud of our stewardship of these lands," the chief executive officer of Crescent, Richard C. Ranson, declared. "Now we have created an opportunity for the legacy to pass to all the people of the Carolinas—to be owned and enjoyed by them for as far out as we can look."[3]

The Associated Enterprises Group contributed $52 million in net income to Duke Power in 1994, which was a 142 percent increase over the 1993 figure. The company projected that the AEG would provide a minimum of $100 million in net income by the end of 1998 and that it would also be a major player in gas-fueled generation. "We are leaving a world where there are safety nets and ceilings," Bill Grigg noted, "and going where we must take risks, but where we can reap great rewards." It was, he believed, "a lot like growing up."[4]

When asked just how big were the international ambitions of the AEG, its head, Bill Coley, pointed out that Duke Engineering & Services was already operating in twenty-five countries on six continents. It saw its market as wherever high-quality engineering and technical services were needed.

Investing in assets was another matter altogether. Through Duke Energy the company was interested in owning and operating power generation and delivery facilities, but the risks had to be carefully assessed. "We look at a country's political stability, the currency risks, the condition of the facilities—everything," Coley explained. Duke Energy was already in Argentina, was interested in Chile, but at that time had no interest in Mexico.

Asia, and especially the Pacific Rim countries, promised to be a big market, and Duke/Fluor-Daniel had already gone into Indonesia. Duke Energy had signed a twenty-five-year contract with Exxon to pursue power investment opportunities in certain parts of China.

As for Eastern Europe, Duke Engineering & Services was doing some engineering work in Ukraine, but the area was not then regarded as a target for investment opportunities.

Coley noted that Duke was not interested in owning any distribution properties outside of the United States. That was because the distribution business was intimately entwined in a country's culture, especially when one considers collecting bills and providing direct customer service. "We believe that's outside our 'core' competencies," Coley explained, "even though we are very willing to do technical and engineering work for others."

There was a pattern of the AEG units working closely with other, well-established companies in pursuing new business and new markets. Duke/Fluor Daniel was an early example of that, for Fluor Daniel, a giant construction company headquartered in California, was a "brand" that was respected around the world. Duke Engineering & Services had teamed up successfully with big names such as Lockheed and others. DukeNet joined forces with Bell South and others with the intention of becoming "the number one provider of personal communications services in the two Carolinas."

Since Duke's newest strategic plan called for the AEG to become a major player in gas-fired generation, what did Coley expect on that front? He pointed out that Duke already owned gas-fired generation in Argentina, but "We want to be involved broadly in the gas market [in the United States]." For one thing, if a company was selling gas-generated electricity, it was good business to establish a position in gas to reduce fuel risks. "If you're working with your customers to lower the energy content of their product," Coley argued, "you have to be able to deal in all forms of energy—electricity, gas, oil, coal."

Nantahala Power & Light, a home-based member of the AEG, faced the same modest 2 percent growth rate as Duke Power. Coley believed, however, that there was potential for some additional growth through economic development, on which he and his associates had been working. Since the United States Forest Service owned much of the land in the Nantahala service area—as much as 80 percent in one county—that limited the potential of development. But it was a desirable area to live in, and there was an available work force with much to offer. "They have a marvelous work ethic in the Nantahala service area," Coley declared, "and we're trying to help them attract the right business."

Coley noted that he had originally seen AEG's most important contribution to Duke Power as employing the utility's strengths to create earnings growth in businesses closely related to the core business. Subsequently, however, Coley said he had come to realize that there was another equally important aspect: through the AEG enterprises "we are acquiring skills that are critical to our success as a utility in the new markets." The electric utilities and those who worked for them sometimes had the reputation of being hidebound, Coley pointed out. But he did not think that was true of Duke Power. "Our people display a real creative and entrepreneurial bent," he stated, "and when you put them in a competitive situation, they catch fire." There were some skills or experience that the AEG units sometimes needed to bring in from the outside, but "in most cases we can find the talent we need inside the company."[5]

As important as the AEG was for Duke's bottom line, the core business—electric supply in the Piedmont Carolinas—remained fundamentally important. Even with Duke's aggressive load management program, keeping ahead of the peak demand remained a constant challenge. In late 1989 Duke announced that it had selected a site in Lincoln County, North Carolina, (northwest of Charlotte) for a combustion turbine facility. Up to 16 turbines were to be installed as needed and could be run on either oil or natural gas during peak periods, which gave Duke the option of choosing the most economical fuel. Combustion turbines could be built relatively quickly and, once installed, could be brought on-line and off-line quickly. Moreover, if natural gas proved to be the chosen fuel (as it did), its use produced significantly fewer pollutants than either coal or oil.

Part of Duke's plan to meet future power needs under a "least cost" strategy, Lincoln Combustion Turbine Station was fully completed in 1996 (though some units went on line earlier), three months ahead of schedule and almost $100 million below budget. The 1,200-megawatt capacity of the station added considerable flexibility in meeting system demand. (A megawatt equals a million watts.) Combustion turbines and natural gas were to play an increasingly significant role in Duke Power's operations.[6]

An actual example of Duke's problem about peak load, and how load management helps in dealing with it, may aid in understanding the matter. The entire Eastern Seaboard coped with extremely cold weather in January, 1994. On the 19th, after several days of low temperatures, William Sands, Duke's manager of energy control, feared a crisis, for some seaboard utilities had already gone to rolling blackouts or voltage reductions. By 4:00 a. m. Sands was already telephoning from his home and worrying about the tight-supply situation.

He was at work in the office by 6:00 a. m., and the temperature kept dropping. "By 7:00 a. m. we knew we were going to have to go to load control," he later reported. And in the hour between 7:00 and 8:00 a. m., Duke's customers set an all-time peak demand (up to that time) of 16,070 megawatts. At 7:30 a. m. the coordinators activated load control to interrupt service to water heaters of participating residential customers. At the same time, they implemented load control with industrial and commercial customers who participated in the standby-generator program. Then at 8:02 a. m. Duke implemented load control with its industrial customers in the interruptible-service program.

"Load control gave us what we needed," Sands boasted. The peak lasted forty minutes before the demand began to drop. He observed, too, that Duke's combustion turbines had been "magnificent," producing even more than their rated capacity.[7]

Duke Power's customers never knew, of course, that their electrical service had been in jeopardy on that frigid morning. In fact, the main occasion for appreciating electricity is when one fails to receive it, as often happens after one of the severe ice storms that occasionally cripple the Carolinas or when fierce hurricanes sometimes waywardly blow into the Piedmont.

Through the course of the century, Duke employees coped with countless emergencies produced by ice storms, tornadoes, and hurricanes, but few were as devastating as Hurricane Hugo in 1989. It hit the mainland in the vicinity of Charleston, South Carolina, on September 21. A storm surge with 135-mile-per-hour winds knocked bridges off pilings, stranded boats in the middle of highways, did severe damage in Charleston, and virtually wiped several small coastal towns off the map.

Although the forecast had been that Hugo would pass to the east of Charlotte, by 3:00 a. m. on the 22nd residents of that inland city were awakened by the sound of gusting winds and cracking trees. For the next three hours, the area was buffeted by torrential rain and 85-mile-an-hour winds. Charlotte had always prided itself on its trees, but as an estimated 80,000 of them cracked and crashed, they wreaked havoc with Duke's transmission system.

The company had prepared for the storm, but the breadth of it and extensiveness of the damage were unprecedented. Until Hugo, Duke's worst storm in recent memory had occurred in May, 1989, when a series of tornadoes swept through a portion of the service area and left about 250,000 customers without service for up to several days, with the heaviest damage in Winston-Salem. That proved to be, however, only a warm-up for Hugo, which robbed nearly 700,000 customers of power. In Charlotte and vicinity, the hardest-hit area, 98 percent of customers (232,000 out of 237,000) were without power and, for some, repairs took over two weeks.

As happens sometimes when disaster strikes, "adversity brought out the best in people," as the company reported. Duke crews experienced the generosity and gratitude of customers, who brought out coffee and food. For many line technicians, Hugo meant weeks away from home, and ultimately 9,000 persons participated in the recovery effort—6,500 of whom were Duke employees and local electrical contractors whose crews frequently worked with the Duke system and some 2,500 who came from as far away as Mississippi, Florida, New Jersey, and Indiana.

Working 16-hour shifts the first week (down to 12 and 14 subsequently), the crews struggled to restore the mangled system. Duke's customers responded gratefully, for in addition to supplying food and drink, members of a church in Charlotte washed, dried and ironed the clothes of one crew. Residents in small towns sponsored dinners at local restaurants in honor of the crews, as did members of some rural churches.

Patience did thin as days without power dragged on, and isolated outages amidst areas where power had been restored inspired sometimes angry reactions. Heavy rains that descended on the Piedmont a week after Hugo did not help matters, for they not only slowed repairs but also flooded some areas. The restoration of Duke's distribution network took a bit over two weeks.

Bill Lee reported that the storm had required the most massive clean-up and repair effort in the company's history. The damage cost the company approximately $64 million; the capital portion of the costs, about $44 million, was capitalized, to be depreciated over approximately 30 years. The company received permission from the appropriate regulatory commissions to amortize for accounting purposes the remaining $20 million over a five-year period.[8]

Hugo was one kind of storm, but winds of another type—from deregulation and competition—kept blowing through the electric utility industry with increasing intensity in the 1990s. One response from Duke Power was to modernize its older coal-burning plants and make them as efficient as possible. As far as fuel-efficiency alone was concerned, Duke had long been at the top of the list. Production costs as a whole, however, included much more than just fuel efficiency, and Rick Priory argued that Duke's coal-fired units ranked about 60th in production costs. "If you're in 60th place you're going to have some struggles in a competitive market place," he warned.

Under a regulated-monopoly system, utilities had not had to worry so much about inefficiencies, for their costs would usually be covered. Now in the 1990s, however, the pressure was on for as low rates as possible, and Duke's rates were lower than a decade earlier. Yet the company had been forced by inflation to pay wage and salary increases and to pay more for supplies—"And we can't go to Raleigh and Columbia and ask for a rate increase."

Through a "business process improvement" program, Duke sought to increase productivity. The program involved having employees cross-trained to gain skills in different areas, changing shift schedules, and reducing overtime hours. Understandably enough, all the changes resulted in some upheaval and dissatisfaction among employees. Admitting that the "business process improvement" program had led to a drop in morale, Priory hoped that "as we get used to doing our work in new ways, I believe the stress will come down."

At the old Dan River Steam Station, a Duke workhorse since 1949, Priory had to combat rumors that the station was to be closed and sold for scrap. On the contrary, he explained that "we're doing all we can to make it competitive in whatever kind of market emerges." There were plants similar to Dan River that could operate annually for $2 million less in operating and maintenance costs, and that was the competition. Duke had to reduce those costs, Priory maintained, as well as take more risks on the spot market for coal in order to reduce fuel costs. "We want to give you every opportunity to run this plant the best it can be run,' Priory advised the disconcerted employees.[9]

And even in the high-tech era of the 1990s, Duke's hydro plants still had an important role to play. There were by then 25 conventional hydro stations producing a bit over 1 million kilowatts and two pumped-storage stations, Jocassee and Bad Creek, that produced 610,000 kilowatts and 1,065,000 kilowatts respectively. A new "Hydrovision Program" involved rebuilding generators, overhauling turbines, replacing much supporting equipment in the plants, and installing computerized touch-screen controls in most of the stations. Rick Miller, the senior engineer in the Hydrovision project, explained that dispatchers still counted on hydro to come on line in ten minutes. When Duke Power hits a peak, he noted, the price of electricity gets high on the open market, but the cost of Duke's hydro generation could be a fraction of a cent per kilowatt hour. That made hydro "critical to our reliability—and our profitability." Miller thought there was a nice symmetry about Hydrovision: "A company that began the 20th century as a hydroelectric utility will begin the 21st century no less committed to bringing out the full potential of its hydro plants."[10]

Amidst the constant swirl of change around and within Duke Power, the specter of downsizing—workforce reduction—would not go away. Asked in the fall of 1994 about the sensitive subject, Bill Grigg replied: "I think we're where we ought to be right now. We don't anticipate any other significant reductions in our work force at this point." While 16,000 seemed about the right number at that time, Grigg added, "You never say never because tomorrow is not going to be like today." Earlier Duke Power had a large construction department with several thousand people in it, but by 1994 there was no construction department. He suggested that the company felt an obligation "to keep folks fully trained and up to speed so that as new opportunities open, folks are qualified to take advantage of them."[11]

Grigg was wise to declare that "one never says never," for a year later, in September, 1995, he announced that management had reluctantly come to see "strong indications that targeted layoffs will be necessary, in various parts of the business, going forward." He continued: "Frankly, the pace of change and competitive pressures have outpaced our capacity to find the full array of job opportunities, in spite of what I consider outstanding efforts on the part of business planners and marketers in both regulated and unregulated business units." The Associated Enterprises Group employed about 1,500 persons, most of whom had come from within Duke Power. But every segment of every market that Duke had entered had opened to strong competitive players. Other utilities were matching Duke's strategies in engineering services, plant construction and management, and energy marketing.

At the same time, Duke's core business in the Piedmont Carolinas now operated in a market where strong downward pressure on prices continued to operate. "All of this is unprecedented in our industry, " Grigg insisted, and "it is truly revolutionary." The old world of utility monopolies was gone forever, and as it disappeared, Duke tried to get a clear focus on "a new world that only comes into view a piece at a time."

Duke Power served a large industrial and commercial market, Grigg explained. That meant that competition was hitting Duke harder than some utilities, for the most competitive arenas in the utility business were already in power generation and in serving large industrial and commercial customers. Utilities like Duke, which benefitted from a strong financial footing could, if they wished, "buy a few precious years of status quo with their financial and human capital." Utilities that went that route, however, might not survive the 1990s.[12]

When asked how workforce reductions played into the possibility of a merger or acquisition for Duke Power, Rick Priory offered a candid (if not necessarily employee-comforting) answer: "Our overall intent is to become more competitive and to grow the electric business. To the extent we are competitive in the marketplace, we become a more attractive merger partner—the more attractive we are, the more opportunities we have." Priory argued that Duke's purpose was "to become the best electric utility in the land," and he insisted that "we are on our way to that goal."[13]

In the final analysis, the simple truth was that deregulation and competition were not old-shoe comfortable for anyone—from top executives to rank-and-file employees—in the electric utility industry of the 1990s. The *Duke Power Journal* put the matter both succinctly and colorfully when it suggested that the industry "used to be like a peaceful river—slow and safe." But all the changes brought on by deregulation and competition had "turned that river into whitewater rapids."[14]

Despite all the uncertainty and discomforting change, Duke Power obviously continued to favorably impress many people in the industry. The company won an unprecedented distinction when the Edison Electric Institute, the industry's influential trade association, announced in 1996 that for the third time Duke Power had been awarded the Edison Award, the industry's highest honor. The panel of judges who made the award cited the following key areas as reasons for Duke's selection: strong financial performance; on-going customer and community partnership; continued operational excellence of generating stations; the formation of the units in the Associated Enterprise Group; and the successful, ahead-of-schedule startup of the Lincoln Combustion Turbine Station.

To accept the award in Dallas, Texas, Bill Grigg took with him five employees. One of them, a woman in Customer Service, later reported, "The whole thing gave me goosebumps, that we had been chosen to be there." A male employee who went to Dallas said: "I've been with this company for 17 years, almost 18 now, and this is the best thing to happen to me that had anything to do with this company. This is the highlight of my career so far." A woman who was a nuclear operations specialist at Oconee Nuclear Station, reported that she had stood with Grigg at the back of the auditorium to watch a video presentation of Duke Power's accomplishments. She continued: "...I asked Mr. Grigg how it felt to lead a company to this much success. He looked at me, and then grinned, and said, 'I just stay out of the way.'"[15]

Grigg's ingratiating modesty notwithstanding, the company's board of directors approved a new incentive program for 1992 and thereafter that gave both executives and employees a way to increase their overall compensation based on the company's performance. For employees, outstanding unit performance that met or exceeded incentive objectives—coupled with good company performance—would earn a cash bonus of up to 3.6 percent of pay.

Under the new executive incentive plan, top management had to achieve the following goals for extra compensation: 1) reach a minimum return on equity of 12.5 percent; 2) hold total electric operation and maintenance costs to no more than 3.17 cents per kilowatt-hour sold; and 3) spend no more than $136.54 in capital per customer equivalent.

Back in 1978, still shaken by the financial crisis of the mid-1970s, Duke Power had boasted about being a no-frills outfit. Among other things mentioned to support that claim had been the fact that the company's chief executive officer then received one of the lowest salaries paid by any major public utility. In the new competitive era of the 1990s, Duke's board of directors decided that such a policy concerning executive compensation was dangerous to the long-range health of the company. Comparison with a group of fifteen national utilities of about Duke's size revealed that Duke's executive officers earned approximately 20 percent less than their counterparts at other companies. In the interest of the company's future, the goal of the board of directors was to bring executive compensation up to market levels. There were to be no bonuses, however, if the company performed poorly.[16]

Enjoying the boost to executive compensation were more women than had been the case a generation earlier. Ruth Shaw, a senior vice president and the chief administrative officer, was named as Charlotte's Businesswoman of the Year for 1995; she carried on the company's tradition of community service by chairing the 1996 campaign for Charlotte's Arts and Science Council and had previously served as president of the Charlotte Rotary Club. Sharon Decker, who became the general manager of the Customer Service Center in 1990, was elevated to a vice presidency in 1992 and named as chief communications officer and vice president for communications and community relations in 1994. Previously a utility regulator in Pennsylvania, Lisa Crutchfield, an African American, became Duke's vice president for energy policy and strategy in 1997, and Roberta Bowman was named as vice president for public affairs. Sue Becht served as treasurer of the company, Ellen Ruff as secretary and deputy general counsel (and both soon became vice presidents), and Barbara Orr as vice president, southern region.

Male African Americans also moved into the ranks of top management. Ron Gibson, a lawyer, was named as vice president for customer planning in 1992. In 1997, Richard Williams, who had been Duke's manager for the Durham District, became vice president for business and community relations; and Wilfred Neal, a certified public accountant, was named as vice president for audit services.

Duke Power also made progress concerning the utilization of minority and women suppliers. From the establishment of the program to encourage that in 1983, it grew by 1993 to a total of $30 million paid to businesses owned by minorities and women, and the company bought from more than 200 minority-owned businesses and 700 owned by women. "Duke Power is committed to doing business with firms that mirror our diverse customer base," Donna Papa, the director of the program, explained.[17]

Duke began what it called the Developmental Partner Program in Charlotte in 1986. It was a two-year curriculum for minorities who were high-school juniors and seniors and consisted of classes taught by employee volunteers in time management, test taking, interview skills, resume writing, and other topics designed to help prepare students for college and a career. In 1991 the program was expanded to Chapel Hill and in 1992 to several other cities in both Carolinas. The program received the Edison Electric Institute's award for affirmative action and the United States Department of Labor's voluntary effort award.[18]

While recognition for the company's efforts along the line of citizenship and service was always welcome, that was admittedly not foremost on the minds of either Duke executives or employees in the mid-1990s. The big, overarching question concerning the future of Duke Power was this: was there to be a merger or acquisition that would dramatically change the company? Related to that was another question: were the company's constant refinements of its strategic plan really doing any good and had Duke Power actually become, in Rick Priory's words, "a more attractive merger partner?"

To that last question, Don Denton, a senior vice president and chief planning officer, had a most positive answer. He began by noting that in the previous few years Duke had moved quickly to adapt to events affecting the industry. It had adopted total quality management, re-engineered the company's work processes, strengthened the focus on customers, and created the Associated Enterprises Group to add value from non-regulated businesses. Duke reshaped its benefits and compensation programs to make them more competitive and tied to performance.

Another major change, Denton pointed out, had come in 1995 when Duke requested proposals (bids) from other power generators to provide capacity to meet anticipated demand in the late 1990s. In doing that, Duke had moved away from its proud tradition of providing power solely by designing, building, and operating its own plants.

Denton maintained that the strategic planning process had played an important role in nearly all of the above-mentioned changes. For example, the company's strategic plan called for Duke Power to be the most efficient and competitive supplier of electricity in the markets it served. Consequently, management, supervisors, and employees developed procedures and policies that significantly reduced costs and allowed Duke to maintain low, stable rates. The company's total cost per kilowatt-hour delivered had declined from 5.83 cents in 1992 to 5.35 cents as of mid-1996. Productivity as measured by total megawatt-hours delivered was up 56 percent since 1989. Duke employees had cut the time needed to refuel nuclear reactors, the company's principal source of base load generation, by about a third, from 60 days to about 40. Denton then noted that the strategic plan had identified the need to increase earnings from Duke's non-regulated businesses, and that led to the formation of the Associated Enterprises Group. Its income was up 105 percent since 1992.

All of those improvements, Denton concluded, had produced positive effects, including an increase in the price of Duke Power's common stock: the most recent 52-week high for the stock had been $53, nearly the same as in the fall of 1990 when Duke declared a two-for-one stock split. That meant that the market value of the company had nearly doubled in six years. So, though Priory and not Denton used the phrase, Duke Power had indeed become "a more attractive merger partner."[19]

Talk about a possible merger or acquisition for Duke Power became widespread within the company by the early 1990s. At a meeting with a group of employees from across the service area in August, 1993, Bill Lee got several questions from the audience about mergers and acquisitions. "I predict you'll see a flurry" of them in the next several years, Lee replied, "driven by competitive pressure to reduce costs." Duke would be actively looking and considering, but a merger or acquisition would have to meet two tests: "It must be good for our customers and good for our investors." For example, Lee pointed out that Duke's purchase of a utility with significantly higher rates could prove not to be in the best interest of Duke's customers if the acquisition meant higher rates for them.[20]

In another venue, two employees in the power generation group asked Lee if, in light of new development in the natural gas industry, Duke should consider merger and acquisition options in that industry. Lee endorsed the proposition that utilities that owned both gas and electric services might well have an advantage in the rapidly changing market. Beyond that, however, he would only say that Duke Power was "committed to be[ing] a company that will compete and win in a more competitive environment."[21]

After succeeding Lee as chief executive officer in 1994, Bill Grigg continued to get frequent questions about any plans Duke might have for acquiring or merging with another company. In May, 1995, Grigg declared that no negotiations with another company were underway at that time.

"We believe, however, that the changes that are taking place in our industry," Grigg continued, "will result in consolidations, with the stronger companies acquiring the weaker ones." Duke's objective, he declared, was to put itself in a position to become the acquirer if an opportunity arose.

Moreover, Grigg insisted that, in case of an acquisition or merger, some things were non-negotiable from Duke Power's standpoint: "The consolidated company would reflect the Duke Power culture, Duke Power management—the Duke Power way of doing things. Second, in order for a consolidation to be successful, everybody has to win—our customers, the acquired company's customers, and the shareholders of both companies. Obviously, that is not always easy to accomplish."[22]

Not long after that emphatic and clarifying statement, an employee in the company's David Nabow Library put a more pointed question to Grigg. The employee had been reading online news about the electric and gas industries and had noticed that some utilities were beginning to operate in both camps. Had Duke ever thought of getting into the natural gas business to "be 'part of the competition' so to speak?" Grigg replied by declaring that there were several reasons why Duke certainly should consider looking into natural gas. 1) More and more commercial and industrial customers were interested in energy services as a whole and wanted Duke to help them manage their energy consumption to give them every competitive advantage in their own markets. To the extent that their needs might include gas, it would be to Duke's advantage to offer that.

Secondly, Grigg explained that in the 1990s natural gas had become "the fuel of choice for new generation—not coal, not nuclear." Therefore it was desirable for any company that wanted to be a part of that market to develop or have expertise in the area of gas-fired generation.

One of the wild cards in both gas and electricity markets, Grigg added, was bound to be changing technologies. Both gas and electricity providers, and the industries that manufactured their supporting technologies, were working assiduously to make inroads into each other's market. And the markets themselves could change dramatically. Back in the 1970s the Federal government prohibited the use of natural gas in new electric generating plants because of the (mistaken) belief that the supply of natural gas would be exhausted in the near future. In the 1990s, however, after deregulation of the natural gas industry, gas-fired generation was beginning to dominate the market.

"You can be sure of this," Grigg concluded, "Duke Power will not stand idly by and let any changes or competition erode our competitive position."[23]

Sharing the increasingly widespread belief among knowledgeable observers that the electric and natural gas industries were converging, Duke's board of directors authorized Grigg to proceed—cautiously, of course—to explore the possibilities. And that Grigg began to do in July, 1996.

The company that Duke Power set out to team up with was PanEnergy Corporation headquarted in Houston, Texas. One of the largest natural gas companies in the nation, PanEnergy operated over 37,000 miles of gas pipeline, crossing the service areas of some 25 electric utility systems and having a physical presence in 37 states as well as in parts of Canada. Having gone quite successfully through the deregulation of the natural gas industry, PanEnergy had enjoyed strong earnings growth of over 20 percent throughout the 1990s. It was a smaller company than Duke Power in terms of total assets ($7.63 billion as compared with $13.36 billion); net income ($304 million as compared with $715 million); market value ($6.4 billion as compared with $9.72 billion); and number of employees (approximately 5,000 as compared with approximately 17,000).[24] But, like Duke Power, PanEnergy had developed a strong entrepreneurial bent and shared Duke's aggressive, optimistic attitude about the future.

At a meeting in July, 1996, of the board of directors of a company that insured electric and natural gas utilities, Grigg mentioned that Duke Power might be interested in the possibility of joining forces with PanEnergy to a former vice chairman of that company. That encounter started the ball rolling, for it turned out that the Houston-based company was also on the lookout for a partner. As Grigg later told a reporter for the *Wall Street Journal*, he then proceeded to "court" PanEnergy during the next several months, meeting personally with the company's executives 8 to 10 times. It took some months to put the deal together, Grigg explained, because "my mother taught me not to kiss on the first date, [and] it took a while to romance them."[25]

Sticking by the "non-negotiable" matters that he had earlier announced, Grigg won agreement that the new, enlarged company should be called Duke Energy; that the headquarters would be in Charlotte; and that, since Grigg was slated for retirement when he reached 65 in November, 1997, Rick Priory would become the first chief executive officer of the new company.

The last matter was the stickiest point of all, because Paul Anderson, PanEnergy's chief executive officer, was not only a highly successful and hard-driving leader but also, at age 51 in 1996, a year older that Priory. Grigg held firm, however, and Duke Power did, after all, pay a premium to get what it wanted. With Priory as chief executive officer and chairman of the board, Anderson agreed to become the president and chief operating officer of the new company.[26]

The new integrated energy company, the nation's largest, would have a total market capitalization of approximately $23 billion. Under the agreement, which the boards of directors of both companies approved on November 25, 1996, each PanEnergy share was to be converted into 1.0444 shares of Duke Power. Based upon Duke Power's closing price of $47.875 on November 22, 1996, Duke Power was slated to issue approximately $7.7 billion in stock to PanEnergy shareholders to complete the transition. Pan Energy stockholders would own approximately 44 percent of the common stock of the new company. Various state and Federal regulatory agencies had to approve the deal before it finally became official on June 18, 1997—and the Duke Power Company then became a subsidiary of the new Duke Energy Corporation.

While Wall Street frequently looks askance at many large-scale mergers, that of Duke Power and PanEnergy received not only favorable but even enthusiastic notices. Edward Tirello, Jr., a leading utility analyst, declared: "I think you'll find that not only will this be the premier energy company going forward, but I think it will be perceived over the next five years as one of the smartest deals created." *Standard & Poor's* asserted that "Duke is ahead of the curve in establishing its market niche... [and] the company is expected to be a strong and tough competitor." And Morgan Stanley Dean Witter announced its belief that "the whole new company over time will, in investor's minds, equal more than the sum of its parts."[27]

Fortune magazine published a revealing, even if breezily written, analysis of Duke Power as it stood on the brink of becoming a subsidiary of Duke Energy in the spring of 1997. The article opened with the declaration that the electric utility industry was "undergoing perhaps the most exciting, scary, total, overwhelming transformation of any industry in the world. Moreover, *Fortune* believed that no company was coping better with the hurly-burly of change than Duke Power, "the No. 1 U. S. utility in customer satisfaction...."

Utility deregulation, according to *Fortune*, was a three-ring circus. First, there was the wholesale market, which Congress in 1992 opened to competition, allowing anyone who bought kilowatts for resale—12 percent of the national market—to buy power from anyone and requiring utilities like Duke Power to open their transmission lines to "let the juice through."

The same law permitted the states to open their retail markets, "the second ring of the circus, featuring 50 state legislatures complete with clowns, acrobats, lion tamers, contortionists, and lobbyists."

The third ring was the rest of the world, where a combination of privatization and deregulation had made it possible for American and any other utilities to invest and operate globally.

Bill Grigg informed the author of *Fortune's* article that it was his belief that when the game was ultimately played out, the old regulated, local, vertically integrated utility business would be deregulated, global, and disaggregated into five different businesses. One would be power generation, which, since electricity is a commodity, would be a game of costs. The second would be bulk transmission, a natural monopoly that would probably remain regulated. Local distribution would be a third business and a competitive free-for-all.

Business number 4, energy services, would compete on price, reliability, and value added as once-staid utilities offered all manner of new services and even tried to build brand recognition. And fifth—flowing over, between, and among the above four industries—brokers, futures traders, and others would trade electricity and electricity futures contracts in what could become the world's largest commodity market. Such a market was already operating in Argentina, where buyers made bids and traders asked for power on an hourly basis.

While all those things were going on, the *Fortune* piece suggested that utilities could not forget the traditional business that provided the cash without which they could not get in the hunt. "Lesson No. 1, to judge from what Grigg and Duke Power have done," the *Fortune* article suggested, was that "you can't get into the fray unless you also stay above it." That is, Duke Power's strategy, while wide ranging, seemed focused in comparison with that of a number of companies, such as AT&T and Time Warner, that "seem to have become almost incoherent" from the diversification "afforded by a suddenly wider world of opportunities."

The first part of Duke's strategy, as *Fortune* perceived the matter, was "to guard the hen house: 'to operate the most cost-effective, customer-focused utility in the Piedmont area.'" It was an old lesson, perhaps, but one easily forgotten in the giddy euphoria of change: one should secure one's base. To that end, Duke Power had trimmed its workforce nearly 30 percent since 1987; increased productivity (measured in kilowatt-hours per employee) 8 percent a year for eight straight years; and cut inventories 6 percent annually over the same period. In real terms, Duke Power's residential rates were 20 percent lower than they were in the 1980s, "which makes would-be foxes think twice." On top of that, Duke's state-of-the-art Customer Service Center and its special account managers for major industrial customers gave the core business (or "hen house") additional security.

The second part of Duke's strategy had been to build up its energy-services business. Therein, *Fortune* espied another lesson: decide where to place the biggest bets and stick with the decision. Dramatic change created so many opportunities—and threats—that it was easy to let opportunism get the better of focus.

The biggest bet that Duke Power had made so far, *Fortune* declared, was to plan to merge with PanEnergy. "Far from being a bolt-on," the article continued, "the merger will allow the company to offer one-stop shopping as the industry moves from selling products (gas vs. oil vs. electricity) to providing solutions to customers' energy needs, regardless of fuel."

That strategy demanded a set of skills foreign to utilities, the article argued. One was marketing, where PanEnergy was already strong. But Grigg and Duke Power had just brought in a new marketing director, Emmy Lou Burchette. *Fortune* thought it a "revealing choice," for Burchette had not come from a marketing powerhouse like Procter & Gamble but from banking. She had headed marketing for First Union National Bank in Charlotte, and Grigg believed that banking, like telecommunications, might provide a model for how the electric utility business would change. That is, Burchette did not just know marketing; she knew how to invent marketing where little or none had existed.

The third element of Duke Power's strategy was a global manifestation of the first two: to become a major power generator worldwide. According to Bill Grigg, Duke was interested in power generation overseas, not distribution, and was focused on unregulated markets where it could "exploit Duke's traditional strength, which is that for 20 years we've operated the most fuel-efficient coal-fired plants in the U. S."

From that, *Fortune* deduced another lesson: when entering a new market, a company should go in with the business that it knows best. The flip side of that was true also: when entering a new business, a company should begin in the market it knows best. That was why DukeNet, the company's foray into wireless communications, using the technology of the new personal communications services, was centered in the Carolinas.

All of the *Fortune* writer's discussion with Grigg about change inevitably led to talk about the company's culture. Here the writer found Grigg's comment surprising. "We've had all the programs du jour to emphasize competition," Grigg declared. "We've got incentive pay for everyone in the power company, we've done a lot of training, we're forcing competitiveness by eliminating fat."

All those things mattered, but the mindset that Grigg seemed most determined to change had a harder edge: "We have to make capital decisions in an entirely different way." In a regulated environment, with cost-based pricing, there were rewards for adding assets. In an unregulated environment, on the other hand, one could not be certain of recovering an investment, let alone getting a return on it. The rewards came from leveraging assets to raise shareholder value. Therefore, Grigg emphasized that the allocation of capital had to become everyone's concern. For example, employees had to learn to ask themselves whether it was better to buy a backup motor to assure a supply of power or to get the same result by striking a deal with a neighboring generating company. Employees had to learn to identify and sell Duke's intellectual assets and its expertise, quite apart from the kilowatt-hours produced by the physical assets the company owned.[28]

Thus, the widespread reaction to the climax of change for Duke Power was highly favorable. As it and PanEnergy merged to become Duke Energy Corporation in June, 1997, the prognosis looked quite favorable. Only a fool, however, would predict dogmatically about the future, especially in an industry—and a world—where unforeseen changes could suddenly wreak havoc with well-laid plans.

Perhaps a more interesting subject for speculation might be what the principal founders of Duke Power—Dr. W. Gill Wylie, William S. Lee, and James B. Duke would have thought about the company some 93 years after its founding. As for Wylie and Lee, one can only imagine that they would be wide-eyed with amazement. James B. Duke, on the other hand, had known all about hard-driving competition and global markets long before he became involved in hydroelectric power. One has to suspect that if he had been around in 1997, he, after surveying with great satisfaction the extensive industrialization that had so transformed the Piedmont Carolinas during the Twentieth century, would have been raring to get into the frenzied fray of the new era for the power business.

A NOTE ON SOURCES

Upon learning that I was about to tackle a history of Duke Power, my colleague Alex Roland, the department chair and a historian of technology, said that he had just the right book for me: Thomas P. Hughes, *Networks of Power: Electrification in Western Society, 1880–1930* (Baltimore: The Johns Hopkins University Press, 1983). Roland was correct, for Hughes' book is an excellent introduction to a large, complex subject by a leading historian of technology. I found the book most helpful in providing a larger context for the Duke Power story.

On a less ambitious scale than Hughes's book, Richard B. DuBoff, "The Introduction of Electric Power in American Manufacturing," *Economic History Review*, XX (December, 1967) is nevertheless a seminal article on a large subject relevant to this study. Likewise, David E. Nye, *Electrifying America: Social Meanings of a New Technology* (Cambridge, Massachusetts: MIT Press, 1990) proved helpful.

By including a number of primary documents, Beth Ann Klosky, *Six Miles That Changed the Course of the South: The Story of the Electric City,* Anderson, South Carolina (Anderson: Electric City Centennial Committee, 1995) provided me with valuable data for the first chapter of this book. The inclusion of primary sources is a fine idea for such local historians.

Leonard S. Hyman, *America's Electric Utilities: Past, Present and Future*, Fifth Edition (Arlington, Virginia: Public Utilities Reports, 1994) is more of a comprehensive handbook than a history. Described on the back cover as "one of the deans of electric security analysts," Hyman is a widely recognized expert in the field of public utilities, and his book, while not an easy read, proved helpful in various ways.

While various newspapers (especially the *Charlotte Observer* and the Raleigh *News and Observer*) as well as business publications are also cited throughout this study, it is based primarily on a vast and rich collection of material in the Duke Power Company (DPC) Archives in Charlotte. Although a large number of documents from the earliest days of the company had been carefully preserved, it was not until 1986 that Duke Power employed a professionally trained archivist, Dennis Lawson. He promptly set about establishing a first-rate archival and records-maintenance program.

A great deal of the material in the DPC Archives consists of highly technical engineering and scientific data that was of limited value for my purposes. On the other hand, an archival segment known as the "Central File" proved invaluable.

Perhaps the single most accessible treasure-trove in the archives turned out to be something that Lawson prepared and labeled a "Duke Power Archives Chronology." A 981-page computer printout of alphabetized data from the archives, it is actually a wonderfully useful encyclopedia about the company and not really a "chronology." Regardless, it greatly facilitated and sped up my work.

With the great majority of Duke Power stock long held by the Duke Endowment (and, to a lesser extent, the Doris Duke Trust), the company did not begin to publish a conventional, full-fledged *Annual Report* until 1958. From that year forward, therefore, there is important information available from that source. Like all reports that are merely annual, however, the chronological perspective is limited, and one must be alert for public-relations "spin."

The monthly *Duke Power Magazine* began in the 1950s with a small-page format and contained much news about company personnel. Then in 1962 it shifted to full-page format with glossy cover and began to include substantial and informative articles on a wide variety of subjects relating to Duke Power. The company's financial difficulties in the mid-1970s, however, forced the elimination of the magazine.

In February, 1975, the monthly *Duke Power News* began to appear. Less elaborate and expensive than the magazine, the *Duke Power News* was published primarily for the company's employees. Nevertheless, it often carried informative stories about Duke Power. Winning repeated recognition as a top newspaper in the industry, the *Duke Power News* afforded employees an opportunity to question and frequently debate with management. Moreover, as deregulation and competition appeared on the horizon in the late 1980s, Duke Power began downsizing and making a wide variety of changes. When employee polls and questionnaires revealed the growing uneasiness about the changes occurring not only at Duke Power but in the whole, once-staid electric utility industry, the *Duke Power News* (later renamed the *Duke Power Journal*) printed full accounts. It turned out to be a most valuable source.

An oral history conference in Charlotte on December 12, 1997, brought together ten or so Duke Power executives, mostly retired. A court reporter recorded and then transcribed the discussion, and the 141-page transcript of "The 'Old Rats' Meeting," now in the Duke Power Archives, proved to be both colorful and informative.

Likewise, Joe Maynor's *Duke Power: The First Seventy-five Years* (privately published by the Duke Power Company, 1980), 180 pp., is an expanded version of articles that Maynor wrote for the *Duke Power News* to commemorate the company's seventy-fifth anniversary in 1979. The lively, readable text is complemented by a rich sampling from the company's vast collection of photographs.

Notes

Notes for Chapter I

1. Paul K. Conkin, "Hot, Humid, and Sad," *Journal of Southern History, LXIV (February,* 1998), p. 11.

2. This gloomy description of the South around 1900 can be corroborated in many studies, but two of the more recent ones are William J. Cooper, Jr., and Thomas E. Terrill, *The American South: A History* (New York: Alfred A. Kinopf, 1990) and John B. Boles, *The South Through Time: A History of an American Region* (Englewood Cliffs, N. J.: Prentice Hall, 1995).

3. Richard B. DuBoff, "The Introduction of Electric Power in American Manufacturing," *Economic History Review*, XX (December, 1967), p. 512. Herein after cited as DuBoff, "The Introduction of Electric Power..." As the above reference to horsepower suggests, in the early period of industrial electrification people tended to cling to the older terminology. One and a third horsepower equals one kilowatt.

4. Thomas P. Hughes, *Networks of Power: Electrification in Western Society, 1880–1930* (Baltimore: Johns Hopkins University Press, 1983), p. 371. Hereinafter cited as Hughes, *Networks of Power.*

5. Jack Riley, *Carolina Power and Light Company, 1908–1958* (Raleigh, North Carolina: privately published, 1958), p. 53. Hereinafter cited as Riley, *Carolina Power and Light Company.*

6. Beth Ann Klosky, *Six Miles That Changed the Course of the South*: *The Story of the Electric City, Anderson, South Carolina* (Anderson: Electric City Centennial Committee, 1995), p. 87. Hereinafter cited as Klosky, *Electric City.*

7. Leonard S. Hyman, *America's Electric Utilities: Past, Present and Future*, Fifth Edition (Arlington, Virginia: Public Utilities Reports, 199?), p. 82.

8. Klosky, *Electric City*, p. 111.

9. *Anderson Intelligencer*, June 19, 1895, as reprinted in Klosky, *Electric City*, p. 100.

10. Klosky, *Electric City*, p. 111.

11. Undated photo copy of an article about W. S. Lee by W.O. Saunders in *The American Magazine*, RG 34-01-01.3298 in the Duke Power Company Archives. Hereinafter cited as DPC Archives.

12. Klosky, *Electric City*, p. 111.

13. Hughes, *Networks of Power*, p. 1.

14. An older name for the site was Indian Bend, but I have chosen to stick with the name that was in widespread usage by 1900.

15. W. Gill Wylie, Jr.'s recollection in Duke Power Archives Chronology, Key 4900; hereinafter cited as DPA Chronology. Photostat of undated (December 28?, 1912) newspaper article containing a speech by W. Gill Wylie, RG 34-01-01.1123 in DPC Archives.

This speech is also reprinted in Joe Maynor, *Duke Power: The First Seventy-Five Years* (privately published, 1980), pp. 11–13.

16. Ibid.

17. Ibid.

18. Article on W. S. Lee by W. O. Saunders in *The American Magazine*, n. d. (early 1920s?), RG 34-01-01.3298, DPC Archives.

19. Minutes of the board reprinted in Klosky, *Electric City*, p. 105.

20. Ibid., p. 107.

21. Photostat of undated (December 28(?), 1912) newspaper article containing a speech by W. Gill Wylie, RG 34-01-01.1123 in DPC Archives.

22. As quoted under Catawba River Development, Key 4994, DPA Chronology.

23. W. S. Lee, Jr., to W. G. Wylie, June 1, 1904, RG 41-1-1.0363, DPC Archives.

24. W. S. Lee, Jr., to W. G. Wylie, June 1, 1905, RG 41-1-1.0368, DPC Archives.

25. Article on W. S. Lee in *The American Magazine*, n. d. (early 1920s?) RG 34-01-01.3298 DPC Archives.

26. W. A. Erwin to B. N. Duke, November 13, 1899, and Erwin to J. E. Stagg, August 1, 1901, enclosing memorandum of cost of March 12, 1901, B. N. Duke Papers, Special Collections, Perkins Library, Duke University. Hereinafter cited as B. N. Duke Papers. Erwin obtained additional properties at Great Falls which brought the total cost to $90,000.

27. W. A. Erwin to B. N. Duke, April 1, 1905, ibid.

28. W. G. Wylie to B. N. Duke, June 3, October 31, 1899, B. N. Duke Papers.

29. W. G. Wylie to B. N. Duke, May 21, 1902, and Wallace Deane to B. N. Duke, July 29, September 3, 1902, ibid.

30. Transcript of oral interview by Frank Rounds with Roy A. Hunt of the Aluminum Company of America concerning J. B. Duke's visit to Canada in 1925 with various Alcoa officials, in the Duke Endowment Papers, Special Collections, Perkins Library, Duke University. Hereinafter cited as Rounds interview.

31. John Wilber Jenkins, *James B. Duke: Master Builder* (New York: George H. Doran Company, 1927), pp. 173–174. Although Jenkins' biography was authorized by close associates of J. B. Duke and is not a critical or scholarly study, Jenkins had the advantage of interviewing numerous persons who knew and worked closely with Duke. Hereinafter cited as Jenkins, *J. B. Duke.*

32. Ibid., pp. 176–177

33. R. B. Arrington to B. N. Duke, March 7, 1906, B. N. Duke Papers. The Duke brothers each loaned Dr. Gill Wylie a bit more than $118,000 so that he might purchase additional stock, and in October, 1906, the capital stock was increased to $10,000,000.

34. "The Plant at Great Falls," Key 2765, DPA Chronology.

35. Ibid.

36. C. A. Mees and J. H. Roddy, "The Great Falls Station...," *The Engineering Record* (May, 1907), p. 652, as quoted in Key 2372, DPA Chronology.

37. *Charlotte Observer*, April 5, 1908, as quoted in Key 2324, DPA Chronology.

38. Rounds interview with John Fox, June, 1963, p. 18.

39. R. B. Arrington (executive secretary to J. B. Duke) to J. W. Cannon, November 5, 10, 1908, J. B. Duke letterbook, Special Collections, Perkins Library, Duke University.

40. Rounds interview with C. A. Cannon, July, 1963, p. 7.

41. DuBoff, "The Introduction of Electric Power...," pp. 509–518.

42. George A. Gray, "A Visit to Great Falls," *Charlotte Daily Observer*, April 5, 1908, as reprinted in Key 2325, DPA Chronology.

43. Quotation is in James G. Brittain, ed., *Turning Points in American Electrical History* (New York: IEEE [Institute for Electrical and Electronic Engineer's], 1976), p. 162.

44. "The Great Southern Transmission Network," *Electrical World*, 63 (May 30, 1914), as reprinted in ibid., p. 163.

45. Wade H. Wright, *History of the Georgia Power Company, 1855–1956* (Atlanta: Georgia Power Company, 1957), pp. 135–137.

46. "Duke Power Company Innovations," Key 4440 DPA Chronology.

47. *Duke Power Magazine*, May 1948, p. 1, as quoted in Key 1335 and 1488, in DPA Chronology.

48. Robert F. Durden, *The Dukes of Durham*, 1865–1929 (Durham: Duke University Press, 1975), p. 73. Hereafter cited as Durden, *The Dukes.*

49. Ibid., pp. 77–80.

50. Rounds interview with N. Cocke, 1963, pp. 19–20.

51. B. N. Duke to J. B. Duke, July 2, 1913, B. N. Duke letterbook.

52. Hyman, *America's Electric Utilities*, pp. 100–102.

53. Forrest McDonald as quoted in James F. Crist, *They Electrified the South: The Story of the Southern Electrical System* (privately printed, 1981), p. 68.

54. Ibid., p. 54. Wendell Willkie, later to become the unsuccessful Republican presidential nominee in 1940, became the head of Commonwealth and Southern in 1932—before the New Deal's Holding Company Act of 1935 ultimately forced the dissolution of the original company.

55. Erwin H. Will, "...A Story of Virginia Electric and Power Company," pamphlet published by the Newcomen Society in North America, 1965, pp. 6, 13; Riley, *Carolina Power & Light Company*, pp. 249 ff.

56. Durden, *The Dukes*, p. 189.

57. Jno. A. Law, Spartanburg, South Carolina, to J. B. Duke, October 21, 1912, and Duke to Law, October 23, 1912, RG 34-01-01.2090, DP Archives.

Notes for Chapter II

1. Manuscript of speech by C. S. Reed, "Progress in the Piedmont Carolinas," 1954, in the Duke Endowment MSS, Perkins Library, Duke University.

2. David E. Nye, *Electrifying America: Social Meanings of a New Technology* (Cambridge, Massachusetts: MIT Press, 1990, pp. 5–6. Hereinafter cited as Nye, *Electrifying America.*

3. Ibid., p.26.

4. The percentages are from ibid., p. 4.

5. James F. Crist, *They Electrified the South: The Story of the Southern Electrical System* (Privately printed, 1981), p. 46.

6. Key 2767,DPA Chronology.

7. Rounds interview with C. E. Buchanan, August, 1963, p. 12.

8. As quoted in Maynor, *Duke Power*, p. 41.

9. John Fox, outline history of the power company, p. 9, in the Duke Endowment MSS; DPA Chronology, Key 2774 and 2767.

10. Ibid.

11. Rounds interview with Mrs. E. C. Marshall, October 1963, p. 18.

12. W. S. Lee and Richard Pfaehler, "Reservoir and Plant for New Southern Water Power," *Engineering News-Record*, June 6, 1920, pp. 1088–1089, RG 34-01-01.5470. DPC Archives; key 391, DPA Chronology.

13. "Duke Power Electric Service in Piedmont North Carolina," memorandum distributed by Norman Cocke, November 27, 1956, RG 34-01-01-0490, DPC Archives.

14. "To the Commissioner of Internal Revenue, Washington, D. C., February 17, 1927, pp. 13–14, RG 34-01-01.2351, as cited in Key 4972, DPA Chronology.

15. Key 996 and 83, DA Chronology.

16. Key 2812 and 5329, DPA Chronology.

17. *Charlotte Observer*, April 1954 (when Nabow was named a vice-president of the company); Key 628 and 5207, DPA Chronology.

18. Key 5219, and 4392, DPA Chronology.

19. Rounds interview with Edward Williams, April 2963, pp. 19–24.

20. Rounds interview with A. Carl Lee, April and July 1963, p. 49.

21. RG 34-01-01.5466, DPC Archives, and Key 2099, DPA Chronology.

22. The most recent and comprehensive study is David P. Massell, "Amassing Power in a Northern Landscape: J. B. Duke and the Development of the Saguenay River, 1897–1927, a Ph.D. dissertation at Duke University, 1997. Hereinafter cited as Massell, "Amassing Power."

23. John Fox, unpublished outline history of the Duke Power Company, as cited in Massell, "Amassing Power," p. 119.

24. Massell, "Amassing Power," p. 120–121. Massell explains that from 1899 forward, electrochemists on both sides of the Atlantic competed to develop the most effective process to fix nitrogen from the air by using electricity, with the methods differing mainly in the type and design of electric furnace employed. As a chemical engineer of the era explained, when air was heated to very high temperatures, it oxidized to form nitric oxide and, on rapid cooling, remained in that form.

25. Massell, "Amassing Power," pp. 123–124, 297.

26. Jenkins, *J. B. Duke*, p.188. Massell, "Amassing Power," has a much more detailed and nuanced account of all this than Jenkins.

27. W. R. Perkins, "An Address on the Duke Endowment: Its Origin, Nature and Purposes," an address delivered in Lynchburg, Virginia, October 1929 and contained in a printed pamphlet, Duke University Library. For a more detailed account of the origins of the Duke Endowment than will be given in this study, see Robert F. Durden, *Lasting Legacy to the Carolinas: The Duke Endowment, 1924–1994* (Durham: Duke University Press, 1998), Chapter 1.

28. Rounds interview with Norman Cocke, April 1963, p. 185.

29. R. B. Arrington to B. N. Duke, March 25, 1911, B. N. Duke Letterbook.

30. Alex Sands to B. N. Duke, April 2, 1917, and to Mrs. B. N. Duke, August 31, 1917, ibid.

31. Hyman, *America's Electric Utilities*, p. 92.

32. Ibid.

33. *News and Observer*, January 16, 1913, as quoted in Paul Krause, "Implications of the Initial Effort to Regulate the Electric Power Industry in North Carolina," unpublished seminar paper at Duke University, April, 1980, p. 15. Hereinafter cited as Krause, "Implications."

34. Letter of Alfred E. Kahn, "Let's Play Fair with Utility Rates," *Wall Street Journal*, July 25, 1994.

35. Krause, "Implication," p. 21.

36. Ibid.,pp. 26–27.

37. *Charlotte Observer*, December 21, 1919, and *Raleigh News and Observer*, December 21, 1919.

38. *Raleigh News and Observer*, November 23, 1920.

39. *Raleigh News and Observer*, February 25, 1921.

40. Ibid.

41. *Raleigh News and Observer*, February 25, 1921.

42. *Charlotte Observer*, February 26, 1921.

43. *Charlotte Observer*, March 9, 1921.

44. *Raleigh News and Observer*, April 14, 1921.

45. For more details about these matters, see Durden, *The Dukes*, pp. 207–222.

46. *Charlotte Observer*, October 12, 1923; *Raleigh News and Observer*, November 18, 1923.

47. *Raleigh News and Observer*, October 21, 23, November 11, 1923.

48. *Raleigh News and Observer*, October 25, 1923.

49. W. O. Saunders, *Elizabeth City Independent*, as reprinted in *Raleigh News and Observer*, December 23, 1923.

50. *Charlotte Observer*, October 24, 1923.

51. *Natural Resources* editorial, as reprinted in *Charlotte Observer*, November 5, 1923.

52. Raleigh *News and Observer*, November 23, 1923, and January 13, 1924.

53. Corporation Commission;s Order, *Raleigh News and Observer*, January 13, 1924.

54. Key 4968, DPA Chronology

55. Charles P. Roland in his editor's preface to James C. Cobb, *Industrialization and Southern Society, 1877–1984* (Lexington: University of Kentucky Press, 1984), p. ix.

56. Harold U. Faulkner, *The Decline of Laissez Faire* (1951), p. 142, as quoted in Richard L. Wilson, "Cotton Spindles and Kilowatts: A Study of the Cotton Manufacturing Industry and the Southern Power Company," unpublished master's thesis, University of North Carolina at Charlotte (1980), p. 1. RG 34-01-01.4099, DP Archives. Hereinafter cited as Wilson, "Cotton Spindles and Kilowatts."

57. Wilson, "Cotton Spindles and Kilowatts," p. 1.

58. Ibid.

59. George B. Tindall, *The Emergence of the New South, 1913–1945* (Baton Rouge: Louisiana State University Press, 1967), pp. 75–76. Herein after cited as Tindall, *Emergence of New South.*

60. C. Vann Woodward, *Origins of the New South, 1877–1913* (Baton Rouge: Louisiana State University Press, 1951), p. 308. In the following paragraph, Woodward skewers "the master of the tobacco monopoly," J. B. Duke, but nowhere mentions the Southern Power Company or (if the index is correct) even hydroelectricity.

61. Since this matter is dealt with in detail in Robert F. Durden, *Lasting Legacy to the Carolinas: The Duke Endowment, 1924–1994* (Durham: Duke University Press, 1998), it will only be summarized here, with the emphasis on the linkage with the power company.

62. The indenture creating the Duke Endowment is included as an appendix in both *The Dukes of Durham* and *Lasting Legacy to the Carolinas.*

63. J. B. Duke's "Indenture and Deed of Personalty," in Durden, *The Dukes*, pp. 277–278.

64. Ibid., pp. 271–272.

65. Massell, "Amassing Power," pp. 355, 358. The Wateree plant, the largest on the Catawba at that time, was designed to generate a maximum of 90,000 horsepower.

66. Paul Clark, "James Buchanan Duke and the Saguenay Region of Canada," p. 160, privately printed booklet in Special Collections, Duke University Library. Massell, "Amassing Power," has much more detail.

67. Massell, "Amassing Power," p. 368.

68. Ibid., pp. 384–385. Two-thirds of Duke's Alcoa stock went eventually to the Duke Endowment, and after Duke's death the aluminum company in 1926 acquired a controlling interest in the Duke-Price Power Company and its Isle Maligne plant.

69. Rounds interview with Norman Cooke, April, 1963, p. 121.

70. Rounds interview with Roy Hunt in Pittsburgh, October and December, 1963.

71. *Charlotte Observer*, September 13, 1925.

72. Raleigh *News and Observer*, November 14, 1923. By 1930 the system had grown to include a total of twenty-nine companies and was said to be he most extensive interconnection east of the Rocky Mountains. Tindall, *Emergence of the New South*, p. 74.

73. Rounds interview with Mrs. E. C. Marshall, October, 1963, p.35.

74. Key 5218, DPA Chronology.

75. Jenkins, *J. B. Duke*, p. 259.

76. Father Arne Paula, "The Worth of Work," *Wall Street Journal*, September 5, 1998, p. W13.

Notes for Chapter III

1. Robert Dick to Dennis Lawson, February 23, 1989, in RG 34-01-01.5722, DP Archives.

2. C. E. Watkins, "The Construction Department Story," February 24, 1960, in RG 34-01-01.1136, DP Archives.

3. W. S. Lee, speech on Buck Steam Station, n.d. (late 1926?), RG 34-01-1.0283, DPC Archives.

4. Keys 1409, 1931, and 2761, DPA Chronology.

5. Thorndike Saville, "The Power Situation in the Southern Appalachian States," *Manufacturers Record*, April 21, 1927, in RG 34-01-01.4795, DPC Archives.

6. C. 1. Burkholder to G. G. Allen, January 17, 1928, RG 34-01-01.0291, DPC Archives.

7. G. G. Allen to C. 1. Burkholder, January 24, 1928, RG 34-01-01.0291, DPC Archives.

8. Pryor, *Duke Power*, p. 65.

9. Keys 2377 and 4439, DPA Chronology.

10. The literature on this subject, both popular and scholarly, is enormous, but two studies that might serve as introduction are Thomas McCraw, *TVA and the Power Fight, 1933–1939* (Philadelphia: Lippincott, 1971) and Phillip J. Funigiello, *Toward a National Power Policy: The New Deal and the Electric Utilitv Industry, 1933–1941* (Pittsburgh: University of Pittsburgh Press, 1973).

11. Pryor, *Duke Power*, p. 69.

12. W. S. Lee "Water Power Development," *Electrical Engineering* (May 1934,) pp. 715–719. RG 34-01-01.3316, DPC Archives.

13. Minutes of the Trustees of the Duke Endowment and accompanying documents, June 26, July 13, 1934, vol. 7, Box BOT 2, Duke Endowment Archives, Perkins Library, Duke University.

14. *New York Times*, December 8, 1937, pp. 21, 23, and January 4, 1938, p. 1.

15. C. I. Burkholder to (textile-mill owners), May 7, 1930, RG 34-01-01.0289, DPC Archives.

16. Thomas F. Hill, "The Duke [Power] Story," Key 2935, DPA Chronology.

17. Pryor, *Duke Power*, p. 71. The promoting and selling of electrical appliances was actually done by the Southern Public Utilities Company in the early 1930s, but that subsidiary was merged with Duke Power in 1935.

18. Key 1570, DPA Chronology.

19. Nye, *Electrifying America*, p. 287.

20. Tindall, *Emergence of New South*, p. 455.

21. Marvin T. Giddings, "History of the Duke Power Agricultural Engineering Department," typed MS, RG 34-01-01.5459, DPC Archives.

22. Keys 3991, 3988, DPA Chronology.

23. Keys 1202 and 2893, DPA Chronology.

24. Boles, *The South through Time*, p. 455.

25. Ibid., p. 455.

26. Pryor, *Duke Power*, pp. 77–78.

Notes for Chapter IV

1. Hyman, *America's Electric Utilities*, p. 115.

2. Ibid., p. 117.

3. Pryor, *Duke Power*, p. 92.

4. Ibid.

5. Key 3431, DPA Chronology.

6. W. J. (Buck) Wortman, "Duke Power Company—The First Two-Way Radio Installation," RG 34-01-01.5485, DPC Archives.

7. *Wall Street Journal*, March 30, 1998, p. A18.

8. Peter Stoler, *Decline and Fall: The Ailing Nuclear Power Industry* (New York: Dodd, Mead, 1985), p. 16. Hereinafter cited as Stoler, *Decline and Fall.*

9. Ibid., p. 31.

10. "A Brief Chronology of Duke Power Company," pp. 7,8, DP Archives.

11. David Mildenberg, "Flipping the Switch at Duke Power: How Bill Lee uses shock treatment to snap his power company out of its complacency," *Business North Carolina* (September 1991), reprint, p. 1.

12. "A Brief Chronology of Duke Power Company," DP Archives, p. 7.

13. *Duke Power Company Annual Report, 1958*, p. 3.

14. Ibid., p. 7.

15. Ibid., p. 8.

16. Ibid., p. 11.

17. Ibid.

18. Ibid., p. 15. In later years, fuel costs would outstrip taxes.

19. "A Brief Chronology of Duke Power Company," DP Archives, p. 8.

20. "Duke's Legacy," *Forbes*, March 15, 1962, p. 26. RG 34-01-01.5739, DP Archives.

21. *Duke Power Company Annual Report, 1958*, p. 7.

22. "The Crescent Story," RG 34-01-01.6023, DPC Archives.

23. Transcript of Oral History Conference, aka "Old Rats" Meeting, December 12, 1997, Charlotte, North Carolina, DP Archives, pp. 122–128.

24. W. S. Lee to Dennis Lawson, October 9, 1995, key 2133, DPA Chronology.

25. *DPC Annual Report, 1963*, p. 16.

26. *DPC Annual Report*, 1965 p. 4.

27. Ibid., p. 18.

28. "The Crescent Story," RG 34-01-01.6023, DPC Archives.

29. Jackson's Veto of the Bank Bill, July 10, 1832, in Henry S. Commager, ed., *Documents of American History* (New York: Appleton-Century-Crofts, 1949), p. 274.

30. *DPC Annual Report*, 1961, p. 14.

31. Ibid., *1962*, p. 18.

32. Ibid., *1963*, p. 13.

33. Nancy DeWolf Smith, "The Wisdom that Built Hong Kong's Prosperity," *Wall Street Journal*, July 1, 1997.

34. *Duke Power Magazine* (February, 1965), p. 3.

35. Ibid.,pp. 3–4.

36. Ibid., pp. 5–6.

37. *Charlotte Observer*, March 24, 1965.

38. Ibid., March 25, 1965.

39. Ibid., March 26, 1965.

40. Ibid., July 9, 1965.

41. *Duke Power Magazine*, August, 1865, pp., 2–4.

42. Key 4188, DPA Chronology; transcript of Oral History Conference, December 12, 1997, pp. 29–32, DPC Archives.

43. *Duke Power Magazine*, September, 1967, p. 6

44. A Brief Chronology of Duke Power Company, pp. 10–11, DPC Archives.

45. *DPC Annual Report, 1973*, p. 13.

46. *DPC Annual Report, 1975*, p. 4.

47. *Duke Power Magazine*, Summer 1972, p. 1, 5–6.

Notes for Chapter V

1. For a fuller analysis of the Tax Reform Act's impact on the Duke Endowment, see Robert F. Durden, *Lasting Legacy to the Carolinas: The Duke Endowment, 1924–1994* (Durham: Duke University Press, 1998), pp. 183–204. Here the focus will be on Duke Power.

2. Comments of W. B. McGuire and others in the Oral History Conference, December 12, 1997, pp. 64–77 of transcript.

3. The Endowment continued, of course, to be by far the largest holder of Duke Power common stock. Wishing to further diversify the Endowment's portfolio, from the 1970s on the trustees ran into a roadblock: the indenture required the unanimous consent of all fifteen trustees to the sale of any Duke Power stock. One trustee, Doris Duke, held fast to her father's original ideas and refused to go along with the diversification the majority of the Endowment's board had come to regard as urgently desirable. After her death in October, 1993, the trustees moved promptly to arrange for the sale of 16

million of the more than 26 million shares of the Endowment's Duke Power common stock. This matter is dealt with more fully in Durden, *Lasting Legacy*, Chapter 9.

4. Oral History Conference, December 12, 1997, p. 68 of transcript.

5. Ibid.

6. Ibid., pp. 69–70.

7. *Duke Power Magazine*, June, 1970, p. 12.

8. Richard Pierce to D. E. Woodin, Jr., January 31, 1975, in Corporate Communications Department file, RG 34-01-01-1155.02, DPC Archives.

9. Ibid.

10. *Winston-Salem Journal*, April 17, 1970, RG 34-01- 01.1051, DPC Archives.

11. *Greensboro Daily News*, March 14, 1970, RG 34-01- 01.1051, DPC Archives.

12. RG 34-01-01.1050, DPC Archives.

13. DPC *Annual Report, 1969*, p. 22.

14. A Brief Chronology of DPC, p. 13, DPC Archives.

15. *Duke Power Magazine*, Winter 1971, pages after p. 9.

16. *DPC Annual Report, 1970*, p. 30.

17. Richard Pierce to D. E. Woodin, Jr., January 31, 1975, RG 34-01-01-1155.02, DPC Archives.

18. Hyman, *American's Electric Utilities*, p. 142.

19. *Raleigh News and Observer*, December 28, 1997.

20. *Wall Street Journal*, September 24, 1998, pp. A1, A10.

21. DPC *Annual Report, 1973*, p. 7.

22. Ibid., p. 16.

23. A Brief Chronology of Duke Power Company, p. 12, DPC Archives.

24. Carl Horn to the trustees of the Duke Endowment, February 4, 1974, under Public Utilities, Box CF8, Duke Endowment Archives, Perkins Library, Duke University.

25. *Charlotte Observer*, March 31, 1974.

26. Carl Horn, Jr., to the Rev. C. S. Sydnor, Jr., June 20, 1974. RG 34-01-01.1054, DPC Archives.

27. Daniel H. Pollitt, "Hard Times in Harlan County," *Journal of Current Social Issues*, vol. 11 (Spring, 1974), p. 7. RG 34-01-01.1054, DPC Archives. Hereinafter cited as Pollitt, "Hard Times in Harlan County."

28. Carl Horn Jr. to the Trustees of the Duke Endowment, February 4, 1974, under Public Utilities, Box CF8, Duke Endowment Archives.

29. Pollitt, "Hard Times in Harlan County," p. 10.

30. Arnold Miller to Marshall Pickens, January 22, 1974, under Public Utilities, Box CF8, Duke Endowment Archives.

31. *Charlotte Observer*, March 31, 1974.

32. Doris Duke to Arnold Miller, February 7, 1974, under Public Utilities, Box CF8, Duke Endowment Archives.

33. Carl Horn to Doris Duke, April 3, 1974, ibid.

34. *Durham Sun*, October 25, 1973.

35. Duke Power, "Why a Second Booklet," (the company's response to UMW charges), RG 34-01-01.1056, DPC Archives.

36. Ibid.

37. Ibid.

38. Arnold Miller to P. J. Collins, July 5, 1974, under Public Utilities, Box CF8, Duke Endowment Archives.

39. Memorandum from Perkins, Daniels, and McCormack, [John Spuches] to Securities and Exchange Commission, April 18, 1974, ibid.

40. Ibid.

41. United States of America before the N.L.R.B. Division of Judges, Case No. 9-CB-2506, RG 34-01-01.1079, DPC Archives.

42. *Charlotte Observer*, May 26, 1974.

43. Memorandum to author from William Grigg, [ca. December 2, 1998].

44. *Washington Post*, August 30, 1974, RG 34-01-01.1056, DPC Archives.

45. [Richard Pierce?], "History and Reasons for Duke's Entry into Coal Mining," [late 1974?], p. 4. RG 34-01-01.1076, DPC Archives.

46. Untitled memorandum about the Brookside strike, [ca. 1976], RG 34-01-01.1064, DPC Archives.

47. Conversation with Bill Grigg, December 12, 1997.

Notes for Chapter VI

1. *DPC Annual Report, 1974*, p. 3.
2. *Duke Power News*, February 1975, p. 3.
3. *Duke Power News*, February 1976, p. 2.
4. *DPC Annual Report, 1977*, p. 20, and *1978*, p. 6.
5. *Duke Power News*, June 1978, p. 1.
6. Oral History Conference, Charlotte, North Carolina, December 12, 1997, p. 39 of transcript.
7. *DPC Annual Report, 1976*, pp. 8–11.
8. *DPC Annual Report, 1977*, p. 8.
9. *DPC Annual Report, 1979*, p. 13.
10. Ibid., *1980*, p. 16, quoting Dean Witter Reynolds Inc., *Utility Insights*.
11. *DPC Annual Report, 1976*, p. 7.
12. Ibid.
13. *DPC Annual Report, 1977*, p. 8.
14. *Duke Power News*, August 1975, p. 15.
15. *Duke Power News*, April 1985, p. 8.
16. *DPC Annual Report, 1976*, p. 6.
17. *Duke Power News*, February 1976, p. 19.
18. Ibid., August 1976, p. 4.
19. Ibid., July 1977, pp. 4–5.
20. Ibid., April 1978, p. 4.
21. Ibid., June 1979, p. 24.
22. *DPC Annual Report, 1975*, p. 6.
23. Ibid., *1979*, p. 17.
24. *Duke Power News*, February 1975, p. 1.
25. Ibid., March 1979, p. 2.
26. *DPC Annual Report, 1977*, p. 8.
27. Ibid., *1978*, p. 12.
28. Ibid., *1988*, p. 6.

29. Ibid., *1977*, p. 17.

30. *Duke Power News*, February 1976, p. 6.

31. Hyman, *America's Electric Utilities*, p. 43.

32. Henney, memorandum for files, April 26, 1979, with attached report, under Duke Endowment Trustees, Box CF10, Duke Endowment Archives.

33. *DPC Annual Report, 1979*, p. 11.

34. *Duke Power News*, November 1979, p. 3.

35. *DPC Annual Report, 1980,* pp. 13–14.

36. Although the publisher and date of publication are not indicated, Duke Power published the volume *ca.* late 1979.

37. *DPC Annual Report, 1979*, p. 15.

38. *DPC Annual Report, 1978*, p. 3.

39. Ibid., *1981*, p. 3.

40. *Duke Power News*, April 1982, p. 6.

41. Ibid., pp. 7,10.

42. *DPC Annual Report, 1982*, p. 2.

43 Oral History Conference, December 12, 1997, pp. 40–42 of transcript.

44. Ibid.

45. *DPC Annual Report, 1983*, p. 3.

46. Ibid., pp. 9–10.

47. *Forbes*, February 11, 1985, as cited in *DPC Annual Report*, 1984, cover page.

48. *Wall Street Journal*, October 17, 1984, as quoted in ibid., p. 15.

Notes for Chapter VII

1. *DPC Annual Report 1986, p.2.*

2. Ibid., p. 14.

3. Ibid.

4. Ibid.

5. *DPC Annual Report*, 1987, p.12

6. *Duke Power News*, June, 1987, p.7.

7. Ibid.

8. Ibid., p. 17.

9. *Duke Power News*, August, 1975, p. 13.

10. Carl Horn, Jr., *The Duke Power Story, 1904–1973*, pamphlet, address to Newcomen Society, May 23, 1973. RG 34-01-01. 4500, DPC Archives.

11. *Duke Power News*, February, 1975, pp 4–5.

12. Ibid., April, 1975, p. 5. This substantial gift of blood from Duke employees was a recurring news items in subsequent years.

13. Ibid., September, 1981, p. 2.

14. Ibid., April, 1985, p. 8.

15. Ibid., April, 1985, p. 1.

16. *DPC Annual Report, 1982*, p. 3.

17. Ibid., 1985, p. 11.

18. Notes of R. L. Dick, construction department, about executive staff meetings, RG 34-01-01.1142.02, DP Archives.

19. *DPC Annual Report, 1984*, p. 13.

20. *Duke Power News*, August, 1987, p. 2.

21. David Mildenberg, "Flipping the Switch at Duke Power," *Business North Carolina*, September, 1991, p. 7. Hereinafter cited as Mildenberg, "Flipping the Switch. . . ."

22. *DPC Annual Report, 1985*, p. 4.

23. "Speech by Bill Lee at the 1988 PIC Kickoff Rally," July 14, 1988, with R. L. Dick's notes of executive staff meetings, RG 34-01-01.1142.02, DP Archives.

24. Mildenberg, "Flipping the Switch . . . , " p. 7.

25. *Duke Power News*, November, 1988, p. 12.

26. Stoler, *Decline and Fall*, p. 6.

27. Ibid., pp. 177–178.

28. Mildenberg, "Flipping the Switch . . . ," p. 5.

29. *Charlotte Observer*, October 21, 1984.

30. Conversation with Ferrel Guillory, Chapel Hill, November 3, 1998.

31. *Charlotte Observer*, October 21, 1984.

32. Ibid. What Lee hoped for in 1984 became a reality in the 1990s.

33. *Duke Power News*, September, 1991. (Bad Creek Commemorative Issue).

34. Ibid

35. Ibid.

36. Ibid.

37. Ibid.

38. "Questions and Answers on Bad Creek," for internal reference. RG 34-01-01, DPC Archives.

39. Brochure on Bad Creek Hydroelectric Station, n.d. [ca. 1992] in RG 34–01-01.0181, DPC Archives.

40. *Duke Power News*, November, 1986, p. 2.

41. *The Energy Daily*, February 17, 1987, as quoted in *Duke Power News*, April 1987.

42. *Duke Power News*, June, 1987, p. 7.

43. Ibid., October, 1987, p. 2.

44. Ibid., February, 1989, pp. 1–2.

45. Ibid., December, 1988, p. 2.

46. *DPC Annual Report, 1986* , p. 14.

47. *DPC Annual Report, 1991*, p. 4.

48. *Duke Power Journal*, June, 1994, pp. 67. In June, 1992, the *Duke Power News* became the *Duke Power Journal*, which also had a new format. The publication won recognition, in several years, as the best in the industry.

49. Ibid., August, 1996, p. 4.

50. *DPC Annual Report, 1995*, p. 17.

51. Ibid., *1996*, p. 6.

52. *Duke Power News*, August, 1991, p. 6.

53. *Duke Power Journal*, August, 1993, p. 2.

54. Ibid., September, 1994, pp. 6–7.

55. Ibid., November, 1994, p. 6.

56. Ibid., October, 1996, p. 10.

57. Ibid., October, 1995, p. 7.

58. Hyman, *America's Electric Utilities*, p. 147.

59. *DPC Annual Report, 1992*, p. 18.

60. *Duke Power News*, January, 1992, p. 3.
61. *Duke Power Journal*, October, 1994, p. 4.
62. *Duke Power News*, May, 1993, p. 1.

Notes for Chapter VIII

1. *Duke Power Journal*, April, 1994, p. 2.
2. *DPC Annual Report, 1986*, p. 12.
3. *Duke Power Journal*, November, 1996, pp. 6–7. In May, 1999, the North Carolina chapter of The Nature Conservancy informed its members that the bulk of the area, known as the Jocassee Gorges, was to be "protected for the citizens of North Carolina" and that the "dramatic landscape ranks among the most biologically and aesthetically significant natural areas in eastern North America."
4. Ibid., August, 1995, p. 2.
5. Ibid., August, 1995, p. 7.
6. *DPC Annual Report, 1996*, p. 9.
7. *Duke Power Journal*, February, 1994, p. 5.
8. *DPC Annual Report, 1989*, pp. 2, 4, 14–16.
9. *Duke Power Journal*, July, 1995, p. 2.
10. Ibid., August, 1993, p.4.
11. Ibid., September 1994, p.2.
12. Ibid., September, 1995, p. 12.
13. Ibid., October, 1995, p. 2.
14. Ibid., p. 6.13. Ibid., October, 1995, p. 2.
15. Ibid., August, 1995, p. 2.
16. *Duke Power News*, January, 1992, p. 1.
17. *Duke Power Journal*, March, 1994, p. 4.
18. *Duke Power News*, November, 1992, p. 1.
19. *Duke Power Journal*, August, 1996, p. 2.
20. Ibid., August, 1993, p.2.
21. Ibid., July, 1993, p. 2.
22. Ibid., June, 1995, p. 2.
23. Ibid., August, 1995, p. 2.
24. Ibid., December, 1996–January, 1997, special merger insert.
25. *Wall Street Journal*, November 26, 1996, as reprinted in the *Duke Power Journal*, December, 1996–January, 1997, special merger issue.
26. Interview with Bill Grigg, April 12, 1999.
27. *Duke Energy Annual Report, 1997*, has the above quotation and others in a similar vein on the cover.
28. *Fortune*, March 3, 1997, as reprinted in Duke Power Journal, April, 1997, p. 2.

Index